Looking Backward, Looking Forward

FORTY YEARS OF U.S. HUMAN SPACEFLIGHT SYMPOSIUM

8 May 2001

NASA SP-2002-4107

Looking Backward, Looking Forward

FORTY YEARS OF U.S. HUMAN SPACEFLIGHT SYMPOSIUM

Edited by Stephen J. Garber

The NASA History Series
National Aeronautics and Space Administration
Office of External Relations
NASA History Office
Washington, DC
2002

Looking Backward, Looking Forward: Forty Years of U.S. Human
Spaceflight Symposium / edited by Stephen J. Garber.
    p. cm. -- (The NASA history series) (NASA SP-2002-4107)
Includes bibliographical references.
 1. Astronautics--United States--History. 2. Manned space
flight--History. I. Garber, Stephen J. II. Series. III. NASA SP-2002-4107.

 TL789.8.U5 L66 2002
 629.45'009--dc21

                                    2002014550

Preface and
Acknowledgments—Stephen J. Garber

Human spaceflight is the driver for most activities that the National Aeronautics and Space Administration (NASA) undertakes. While NASA certainly has a rich aviation research heritage and has also done pathbreaking scientific and applications work using robotic spacecraft, human spaceflight is a difficult and expensive endeavor that engenders great popular enthusiasm and support for NASA. Much of this public interest stems from pushing boundaries of adventure, by exploring the unique and challenging physical environment of space. Humans can also perform tasks in space that machines cannot. We can think, analyze, and make judgment calls based on experience and intuition in real time.

In little more than forty years, we have gone from thinking, planning, and hoping that humans will enter space to having rotating crews of astronauts and cosmonauts permanently living aboard an International Space Station (ISS). We have moved from the Cold War, which set the historical context for superpower competition in space during the 1960s, to joint ISS missions involving over a dozen cooperating nations.

Not only have humans proved that it is possible to survive in the harsh physical environment of space, but astronauts and cosmonauts have conducted important scientific and engineering feats in space. We've discovered that microgravity is a unique laboratory setting that is potentially useful for scientists in a broad array of disciplines such as pharmacology, materials science, and physics, as well as more obvious fields such as astronomy.

The pool of people who have flown in space has also broadened tremendously in the past forty years. We have moved from a group of seven handpicked men that were trained as military test pilots to men and women of many national and professional backgrounds. Diversity has become an avowed goal of most federal agencies, including NASA, so that people of many ethnicities and personal backgrounds not only fly in space but serve in key roles on the ground. Older astronauts in their sixties and even seventies have flown in space. Beyond pilots and commanders, NASA now trains scientists and engineers as payload specialists to fly in space. Even more than the payload specialists who are not "career astronauts," NASA has tried to bring other civilians, such as teachers, into space. The issues of diversity in general and of civilians in space in particular have ebbed and flowed in importance over time but continue to be relevant. In recent history, the subject of paying tourists in space has come to the forefront.

This obviously relates to the ongoing topic of commercialization. Spaceflight has always been expensive, but in the 1980s, and especially in the 1990s, the federal government began looking at ways to privatize certain space activities. Different individuals in the commercial sector have expressed varying degrees of interest in making human spaceflight a profitable endeavor. While robotic applications satellites such as remote sensing and communications have been significant ventures since the early 1960s, both the government and the private sector have warmed to commercialization of human spaceflight somewhat later. In the mid-1990s, NASA turned over certain key operational activities of the Space Shuttle to the private United Space Alliance in an attempt to lower the government's costs for "routine operations." Recently, NASA has also entered into several high-profile joint agreements with private companies that are interested in conducting specific experiments, selling data, or targeting marketing opportunities in space. Whether these activities will turn out to be profitable or otherwise worthwhile remains to be seen, but commercialization efforts certainly have been an important force in space history.

The history of human spaceflight has also been shaped significantly by technologies that were initially developed outside the aerospace sector. The computer and biotechnology revolutions have had major impacts on how space operations are planned and executed. In addition to exponential increases in computing power, the advent of digital microelectronics has made fly-by-wire technology possible, which in turn has improved safety. Like computers, advances in biotechnology enable new experiments, knowledge, safety, and health in space; space research also synergistically benefits the biotechnology industry.

One technology that thus far has proved elusive is an inexpensive, reliable launch vehicle to improve human access to space. There are many reasons this has proved problematic. Perhaps the first is that escaping Earth's gravitational pull has continued to be an inherently difficult task technically. Secondly, many knowledgeable people would argue that the government has not provided sufficient financial resources to address this problem after the end of the space race and Cold War. While commercialization still looms large in the space context, no private companies have devoted truly significant resources to address this problem because they typically believe that their investment will not be rewarded any time soon, and because they often view such research problems as the government's domain. Some people even contend that the real cost of launching humans into space is the unnecessary redundancy in personnel costs of having a "standing army" to launch spacecraft such as the Space Shuttle. According to this controversial line of thinking, the launch technology per se is not unduly expensive, but we need a different paradigm for ensuring that we can launch people into space with reasonable safety and cost factors.

Safety has always been and always should be a primary concern of any program that puts people in a dangerous environment such as space. Nevertheless, our views of safety have evolved historically. Hopefully we have learned to be more "proactive" in preventing accidents, but what does this mean, and how is this actually implemented in practice? Over the past several decades, a growing body of social science literature on risk assessment and management has emerged, but few scholars have seriously analyzed risk in spaceflight from such a perspective. What qualifies as an acceptable risk for a robotic spaceflight may obviously be totally unacceptable in the human context.

Such a safety debate has played out in the struggle to find an appropriate power source for long-duration human space missions such as a voyage to Mars. While nuclear power in various forms may be acceptable to the majority, although certainly not all sectors, of the general public for deep-space planetary probes and the like, it faces greater opposition for human spaceflight. On the other hand, would it be possible to adopt the safety model of the nuclear submarine? While the technology base may be present to make this technically feasible, public opinion in the

United States has seen nuclear power as inherently risky and controversial. While scientists such as astronaut Franklin Chang-Diaz have undertaken research in exotic forms of power such as ion or plasma propulsion, such technologies are still in the distant future.

While the history of human spaceflight has generally been one of great technological achievement and inspiration, spacefarers have also suffered many disappointments, both in terms of human tragedies and in failing to meet goals we have set for ourselves. Disastrous accidents such as the *Challenger* explosion and the Apollo 1 fire are etched into our collective memories and prod us to take prudent risks and be ever vigilant about safety whenever lives are at stake. At another level, we have been repeatedly frustrated by our inability to achieve aims such as routine, reliable, and inexpensive spaceflight. Why have we failed in these areas? Is our technology base still immature, or are there other political, cultural, and social factors that limit our ability to satisfy our yearning to explore space?

What is our next logical step after the ISS? Should we send humans to Mars? Before we attempt such long-duration missions, we still have much to learn. Even though NASA has now flown humans on Skylab and the ISS, most Shuttle flights are only one to two weeks in duration. We still need to understand more about how microgravity affects human physiology. We know it causes motion sickness in many astronauts before they become acclimated, but researchers still cannot predict which ones will become ill, nor is there a good treatment for this ailment. Microgravity also causes bone density to decrease, which can be reversed by exercise in space, but how much exercise and what kind is best? Will astronauts on interplanetary missions be exposed to excessive amounts of radiation over their long journeys? Since weight is a hindrance to lifting spacecraft to Earth orbit, lead shielding may not be the best solution.

Other, perhaps more subtly vexing challenges for long-duration missions fall into the realm of psychology. For months on end, the crews presumably will be confined to quite small spaces that will make submarines seem roomy. In addition to potential claustrophobia, the crew will certainly be very isolated. Not only will no other humans be anywhere nearby, but audio and visual communications back home will not be in real time, so astronauts will not be able to speak directly to mission control if a problem arises or to their families for personal comfort. While submarine crews and polar expeditions may provide some answers for how to deal with the psychological stress of such journeys, human spaceflight to other planets will clearly present unanticipated challenges precisely because it has not been done before.

Public opinion has also influenced the realm of human spaceflight in ethical dimensions. When should we allocate financial and human resources to space exploration instead of other, more immediate problems such as social welfare, poverty, and healthcare? Our values also play important roles in allocating resources within NASA's budget. We must balance, for example, the knowledge that comes from Earth remote-sensing satellites with the inspirational value of having astronauts take us to new places. Ethics also play into issues such as how much or little we alter the environments we are exploring and studying. At the dawn of the space age, few people gave such ethical debates much thought. Indeed, the space race of the 1960s was won with specific engineering achievements, even if some critics would dismiss them as propagandistic stunts. Without the overriding Cold War driver, however, such ethical concerns will likely take on greater significance in the future.

Even more important than determining whether our technologies and crews are prepared for long-duration spaceflight, spacefarers and their supporters would do well to remember that there must be a fundamental rationale for further human spaceflight. Ideally, it should be concise and easily articulated so that the public can readily understand it. Currently, space advocates are struggling to convince Congress and the public why human exploration is important enough to support with government funds at all. Clearly, NASA's future budgets are unlikely to be as large as they were during the early space race, so planners will need to be thrifty and innovative.

The future is likely to bring other unanticipated challenges. Will the Chinese initiate a serious human spaceflight program of their own? Perhaps the future international political situation will make it advantageous for NASA to cooperate with China. Will another country such as Brazil loft astronauts into orbit in the next forty years? Will space become a new battleground for military conflict, despite many years of international efforts to keep it peaceful? Will the discovery of life, even if unintelligent beings, on another celestial body rally efforts for further human exploration of the solar system, let alone further reaches of the universe?

A confluence of anniversaries made the spring of 2001 a propitious time for reflection on a forty-year record of achievement and on what may lie ahead in the years to come. The fortieth anniversary of Alan Shepard's first spaceflight, the first time an American flew in space, took place on 5 May 2001. The fortieth anniversary of Yuri Gagarin's spaceflight, the first time a human traveled into space and orbited Earth, took place on 12 April 2001. Coincidentally, this date was also the twentieth anniversary of the launch of STS-1, the first Space Shuttle flight. In addition, 25 May was the fortieth anniversary of President John F. Kennedy's famous "urgent needs" speech in which he proposed putting an American on the Moon "before this decade is out," initiating the Apollo Project. Last but not least, the Expedition One crew to the ISS had finished its historic first mission in the spring of 2001.

Thus, the NASA History Office joined efforts with the NASA Office of Policy and Plans and the George Washington University Space Policy Institute to put together a one-day seminar on 8 May 2001 on the history, policy, and plans of human space-flight. The seminar was open to the public and featured the viewpoints of those who have flown in space and also of nonastronaut experts. The speakers were a fairly diverse lot in terms of background and views, but all were accomplished in their fields and gave thought-provoking comments.

The program began with opening and keynote remarks by then-NASA Administrator Daniel Goldin and respected author Charles Murray. An inspiring speaker, Goldin challenged the audience to persevere through the inevitable and the unexpected challenges facing human space exploration. Murray related several moving anecdotes about the Apollo program and how its management techniques stood out.

The first panel focused on the experience of spaceflight and featured an Apollo astronaut, one of the first Shuttle astronauts, a scientist, a commercial payload specialist, and an astronaut trainee who had not flown in space yet. Buzz Aldrin talked about his unusual career path to the Moon and about a future launch vehicle system that enthralls him. T. J. Creamer spoke about the continuity of building on the achievements of others before him and specifically mentioned how the daughter of another panelist, Bob Crippen, was a trainer for his astronaut class. Scientist Mary Ellen Weber discussed how significant microgravity research could be for the average person on Earth and also enthralled listeners with her experience of having to look down from on orbit at incoming meteorites. Charlie Walker, the first astronaut to fly specifically on behalf of a company, covered how NASA could best work together with private industry.

The second panel featured a variety of historical perspectives on the past forty years. The distinguished speakers covered such specific topics as Soviet-American reactions during the space race, the importance of safety, and counterfactual history. The author of a monumental volume on the Soviet space program, Asif Siddiqi, reinforced how the perceptions, misperceptions, actions, and reactions of the U.S. and the U.S.S.R. created the dimensions of the space race. John Logsdon posed a number of "what if" questions to push historians to rethink our assumptions of the causes of key events. Astronaut and manager Fred Gregory discussed how thinking about reliability has shifted from forcing people to demonstrate a specific safety flaw before a launch would be postponed to the current situation, where managers must actively show that it is safe to launch. William Sims Bainbridge revised his arguments about the social and cultural aspects of the "spaceflight revolution."

In the afternoon, another panel looked at the future of human spaceflight. A variety of speakers, from engineers and scientists to a philosopher and a popular author, gave their provocative opinions on the challenges facing human spaceflight. Astrophysicist Neil de Grasse Tyson challenged space buffs to think of a major engineering or scientific project in history that was not begun for at least one of three reasons—national security, economics, or ego gratification. Robert Zubrin, a passionate advocate of human missions to Mars, echoed Frederick Jackson Turner's famous frontier thesis that it is our destiny to explore new worlds. Homer Hickam proposed that one underappreciated reason for human spaceflight is to tap solar power for use on Earth, and he evoked Wernher von Braun in emphasizing the need to explain clearly why space exploration is worth doing at all. Ethicist Laurie Zoloth challenged listeners to consider the moral consequences of human exploration of new places. Jim Garvin engaged the audience by discussing exciting new technologies that could be used to send humans beyond Earth orbit.

Finally, William Shepherd, the commander of the Expedition One crew to the ISS, gave his take on some lessons learned from his personal experience that could be applicable for future human spaceflight missions. Shepherd views the ISS as a stepping stone on the way to Mars and discussed his vision for how such challenging journeys could be accomplished. He points out that not only do we need to develop more powerful and autonomous spacecraft to reach Mars, but we also need to address cultural differences and standardization issues inherent in what he believes will be increased international cooperation. Shepherd also argues for consolidating expertise in a National Space Institute, similar to the military service academies.

Such a seminar is not only a collected work in the sense of many authors, but also in the sense of many producers. Many of the same people who helped stage the seminar also helped with the production of this volume. Louise Alstork, Nadine Andreassen, Jennifer Davis, Colin Fries, Mark Kahn, Roger Launius, and Jane Odom of the NASA History Office helped greatly with both the seminar and the book. Jonathan Krezel, Becky Ramsey, and Michelle Treistman of the George Washington University Space Policy Institute assisted John Logsdon in staging the seminar. Many thanks also go to Tawana Cleary, who graciously handled the astronauts' appearances, and to the good folks at NASA TV for all of their work. Special thanks go to Mike Green, of the former Office of Policy and Plans, who helped initiate and organize the seminar, and who also chaired a panel. In terms of producing the book, special recognition goes to Michelle Cheston, Dave Dixon, Melissa Kennedy, and Jeffrey McLean in the Printing and Design group at NASA Headquarters. Thanks to all of these professionals for their help with logistical matters and for stimulating new and provocative ideas that promise to maintain interest in and debates on the course of human spaceflight for years to come.

Introduction—John M. Logsdon

Astronaut Alan B. Shepard receives the NASA Distinguished Service Award from President John F. Kennedy in May 1961, days after his history-making Freedom 7 flight. Shepard's wife and mother are on his left, and the other six Mercury astronauts are in the background. NASA Image S67-19572.

Today is an auspicious day for holding this symposium. Today is the fortieth anniversary of the day when Alan Shepard came to Washington after his historic flight. He participated in a parade, addressed a joint session of Congress, and then came to the White House, where President John F. Kennedy gave him a medal.

On that same day, 8 May 1961, Vice President Lyndon Johnson presented President Kennedy with a set of recommendations concerning the future of human spaceflight that contained a historic memorandum signed by NASA Administrator James Webb and Secretary of Defense Robert McNamara.

These recommendations had been developed in the two-and-a-half weeks after Kennedy, on 20 April 1961, had asked the Vice President to carry out a review to identify a "space program which promises dramatic results in which we could win." This set of recommendations led to Kennedy's decision to accelerate the space program, aim at across-the-board space preeminence, and set a lunar-landing goal as the centerpiece of the space program for the 1960s. A decision wasn't made on 8 May 1961, but the decision paper that led to Apollo and all that followed reached the President that day.

To start this celebration of forty years of U.S. human spaceflight, there's no more appropriate person than the ninth Administrator of NASA, Daniel S. Goldin. Dan has served as Administrator longer than any of his eight predecessors and has made remarkable changes in the organization. I think as we look back ten, fifteen, or twenty years from now at his time as Administrator, we'll find that he set NASA on a productive course for the twenty-first century.

Opening Remarks—Daniel S. Goldin

Daniel S. Goldin with a model of the Mars Pathfinder's Sojourner rover.

What a wonderful day it is. We are taking the opportunity this morning to reflect on what it has meant since 1961 to be a spacefaring nation. We are also looking forward to the next forty years of human adventure in space and what it might bring us as a civilization.

While the specifics of what will unfold during the first part of the twenty-first century are not certain—and that's the wonder of the space program—I can say with certainty that the possibilities are boundless. Accordingly, I am both excited about where we have been and where we are going.

Alan Shepard, of course, had become the first American to fly into space during a 15-minute suborbital flight on 5 May 1961, riding a Redstone booster in his Freedom 7 spacecraft. At the ceremony that followed, President Kennedy recognized the courage and sacrifice of all those involved in America's first human spaceflight. The President commented that Shepard's success as the first United States astronaut was an outstanding contribution to the advancement of human knowledge, space technology, and a demonstration of man's capabilities in suborbital flight.

President Kennedy also juxtaposed the very public flight of Alan Shepard with the secrecy of our rival at the time, the Soviet Union: "I also want to pay cognizance to the fact that this flight was made out in the open with all the possibilities of failure, which have been damaging to our country's prestige. Because great risks were taken in that regard, it seems to me that we have some right to claim that this open society of ours, which risked much, gained much."[1]

1. Remarks at the presentation of NASA's Distinguished Service Medal to astronaut Alan B. Shepard on 8 May 1961. *Public Papers of the Presidents of the United States: John F. Kennedy, 1961* (Washington, DC: U.S. Government Printing Office, 1962), p. 366.

President Kennedy's comments about the risks and rewards of spaceflight are just as applicable today as they were on 8 May 1961. In forty years of human spaceflight, we have achieved enormous successes, gained astounding knowledge about our universe and our place in it, and brought untold benefits to the people of the world.

We have learned to survive in the incredibly hostile environment of space. We have landed on the Moon. We have developed a remarkable vehicle, the Space Shuttle, which enables Americans to travel to and from Earth orbit much more readily than any previous launch technology, and we will have a vehicle that will take us not just to low-Earth orbit, but, eventually, we will develop a vehicle to take us out of Earth orbit.

I'm especially pleased to recognize the leadership of Alan Shepard as the first Mercury 7 astronaut to fly to space. He was truly an American hero, and I'm proud to have known him. Not long after I arrived at NASA, Alan met me to tell me that what we were doing at NASA was very important and that he personally wanted to make himself available. He said that he'd do anything that I asked to help accomplish the NASA mission. If I wanted him to testify before Congress, or meet with senior officials, or speak to schoolchildren, or take a trip across the world, he would be happy to do it.

He was an individual who had been the first American to fly in space, as well as an individual who had walked on the Moon. He offered to carry the message of the importance of human spaceflight to the masses because he believed in it so deeply, and he believed in this great nation of ours. Alan Shepard believed that NASA is a representation of the best that America has to offer.

He was enthusiastic about this fact and always shared it at every opportunity. He left us a legacy of excellence that is unmatched.

We need more heroes like Alan Shepard and the other wonderful astronauts who are opening up the cosmos. They are the modern descendants of Lewis and Clark, Richard Byrd, and Charles Lindbergh. They set their sights on the distant horizon of space and the journey to unknown places, bringing back knowledge and understanding. They inspire us with their perseverance. They lead us, as Americans, to a loftier place, and Alan was the first American there.

In some respects, we have come a long way since Alan Shepard flew the tiny Freedom 7 space capsule forty years ago, but, in other ways, we have not yet journeyed so far. Alan would have been the first to say that while the technology has changed, the curiosity of the human mind and the courage of the human heart remain the same.

Those who venture forth into space are a breed apart. Alan Shepard and every other astronaut should not be thought of simply as passengers or visitors in space. They are blazing a pioneering trail that will be followed by others once they have made the way safe. When we make the way safe, we are going to do great things.

As I was preparing these remarks, I thought about the possibilities. We've been locked in Earth orbit for too long, but we are going to break out. There's no doubt in my mind. The seeds are there. This is the anniversary of NASA's forty years of human space exploration, and it represents an important crossroad. As we celebrate it today, we continue to move toward a visionary goal.

In our quest to make what is envisioned real, we test, we build, we launch, we learn, and we fail. Then we start again and never ever worry about the criticism of failure—because in failure we learn, and we start the cycle again.

So we are not only celebrating the past today, but also drawing a demarcation point from which to envision the future yet more wondrous. Let us work together to make it happen. Let us burn into our brains that this civilization is not condemned to live on only one planet.

Let's burn it into our brains that in our lifetimes we will extend the reach of this human species onto other planets and to other bodies in our solar system. Let's build the robots that will leave our solar system to go to other stars and ultimately be followed by people.

I wish that Alan Shepard could have been here with us today. We lost a true pioneer when he passed on in 1998. He liked to say of space exploration, "I know it can be done," "it's important for it to be done," and "I want to do it." His spirit lives on in that quest for our future in space.

I would like to close by dedicating this activity on the past, present, and future of U.S. human spaceflight to the memory of Alan Shepard, the first American hero of the space age and my personal hero. Thank you very much.

Human Spaceflight and American Society:
The Record So Far—Charles Murray

These remarks give me an excuse to revisit a world that Catherine Cox and I had a chance to live in vicariously from 1986 to 1989 when we were researching and writing about Project Apollo. As I thought about it, I realized that actually very few people in this audience have had a chance to live in that world, either vicariously or for real. For most people today, NASA's human spaceflight program is the Shuttle. The NASA you know is an extremely large bureaucracy. The Apollo you know is a historical event.

So to kick off today's presentations, I want to be the "Voice of Christmas Past." If we want to think about what is possible for human spaceflight as part of America's future, it is essential to understand how NASA people understood "possible" during the Apollo era.

It is also important to understand that the way NASA functioned during the Apollo Program was wildly different from the way NASA functions now. In fact—and I say this with all due respect to the current NASA team members who are doing fine work—the race to the Moon was not really a race against the Russians; it was a race to see if we could get to the Moon before NASA became a bureaucracy, and we won. But the lessons of that experience should be ones that we still have at the front of our minds.

First, I would like to provide some perspective on time scale. Think back to 20 July 1990. This was the twenty-first anniversary of the first lunar landing, but that is not why I chose the date. From 20 July 1990 to May 2001 is the same amount of time as from the founding of NASA to the first Moon landing, only eleven years. If you think back to what you were doing on 20 July 1990, it just was not that long ago. So if we think about what infrastructure for human exploration of space existed in 1958, when NASA started, we realize there was virtually none. At that time, there were few buildings, a small staff, and not a glimmer of the equipment that Mercury, Gemini, and Apollo would use. At that time, the largest booster in the U.S. launcher inventory was the Redstone, which was less powerful than the escape tower on the Saturn V. The Space Task Group that was responsible for NASA's early human spaceflight efforts was formed only a few months after NASA itself.

Occasionally I am asked, "How can we get to Mars?" I am tempted to say, "Well, junk the current space program, go down to Langley Air Force Base, put together forty-five people that have no experience whatsoever, give them eleven years, and they will do it." Now that is facetious, but it is how short the period of time was between ground zero and the first Moon landing.

The speed is only symptomatic, however, of the way that NASA functioned during those early years, and I want to go over a few of those characteristics. The first was simply youth. Of the forty-five people who were initial members of the Space Task Group, Robert Gilruth was the oldest at forty-four. Joe Shea and George Low got their jobs at thirty-two and thirty-four, respectively. Chris Kraft got his first big job at the age of thirty-four. Glynn Lunney and Gene Kranz, lead flight directors during the big Apollo missions, became flight directors in their twenties, and they were still barely into their thirties when they were lead flight directors for the Apollo flights.

People were very young, and it made a difference. As you talk to the people of Apollo, they will say over and over, "We didn't know we couldn't do it." People who were older who would try to come into this business often were not able to cut it. The reason they could not cut it was that they were too aware of all the ways that things could go wrong.

One of the things that youth brings with it is an ability to form tightly knit teams, another characteristic of the early NASA. It was so small to begin with that everybody knew one another. Even though by the time Apollo flew, it had mushroomed into tens of thousands of people, those initial connections remained. There were people who had known each other at Langley Center and at Lewis Center who dealt with each other in ways that had nothing to do with their places in the organization charts.

Joe Bobek, who was a second-generation Polish immigrant with only a high school education but a genius mechanic, became chief inspector for the Apollo spacecraft. In contrast, George Low was the courtly offspring of an affluent Austrian family, a brilliant engineer, and exceedingly well educated. Before every Apollo flight, George Low would take a sandwich down to the pad and sit down with his old mechanic buddy from Lewis Research Center. They would talk about what George Low needed to know about that spacecraft.

You had people such as Joe Shea and George Low taking demotions all the time during the Apollo Program. They were sent out of Washington to the Centers. They were technically far lower on the ladder than they had been before, but the reason they did that was because that was where the action was.

I do not want you to feel that I am completely unrealistic and starry-eyed about Apollo. Were there any people who were mostly concerned about their careers during Apollo? Of course there were. But if you talked to a lot of people from Apollo, you also got a very clear message with lots of evidence that this was the period of their life when their personal careers really weren't nearly as important as focusing on the job at hand.

People were calling back and forth, ignoring lines of hierarchy in their quest to solve problems. Incredibly brilliant engineers were running the program. People such as George Mueller, who was in charge of human spaceflight at Headquarters, were extremely well-versed in virtually all details of their programs. In terms of engineering, Mueller could wrestle to the ground a relatively low-level engineer on his own particular specialty. The same thing could be said again and again for people such as Shea, Low, Max Faget, and all the rest. They were managers, yes, but they knew just about everything there was to know about the systems they were dealing with, and this made a lot of difference when they wanted to obtain the respect and the overtime work and the commitment of the troops.

Another important aspect of the program, which you can get away with more easily when it's a young program, was its incredible audacity. I shall give you three examples.

The first example goes back to George Mueller in 1963. He came into NASA as head of human spaceflight and set his underlings to work on a comprehensive look at the schedule and how it was going. They were not going to get to the Moon before 1970 or 1971; that was absolutely clear. So what did George Mueller do? He imposed on the Centers all-up testing. This meant that the first flight of the Saturn rocket, with its mammoth 7.5 million pounds of thrust in the first stage alone, would be with all three stages in that stack. All three stages were untested when Mueller made this decision. This approach was anathema to the German rocket team down at Marshall. The Germans had done very well by testing incrementally, one piece at a time.

The engineers from Langley had done very well taking their experimental aircraft over the years and testing them out one step at a time. Here was this guy from the ICBM world, the third culture, as it were, that made up NASA in those days, telling them, "We're going to do all-up testing—we're going to do it all at one time." No committees made that decision. George Mueller made that decision. It was not a political decision. He was not doing it just to get to the Moon before 1970, although that was clearly one of the motivations for it. But the engineering logic behind it was absolutely fascinating. I recommend you look at this decision-making process as a case study of rigorous engineering thinking combined with enormous willingness to do what was necessary to get a job done.

The second case of audacity was George Low's decision to make Apollo 8 a circumlunar mission. Again, in reconstructing how it was done and why it was done, we are not talking about some wild-eyed adventure. There were engineering reasons why it was possible and why it was not only possible, but valuable. But it was the kind of decision which pushed everything in the schedule a quantum leap ahead of where it would have been otherwise.

The third case of audacity is not a particular event; it is the years that Joe Shea was the head of the Apollo Spacecraft Program Office. It has been Joe Shea's legacy to be remembered as the guy who was pushing so hard that mistakes were not caught, and we had the 1967 fire that killed three astronauts. This was a very controversial period in NASA history, and Joe Shea certainly took the fall for the accident. Nobody was tougher on Joe Shea than Joe Shea was on himself.

I submit to you that he was doing exactly the same thing that George Mueller and George Low were doing. But the coin came up tails for him. But if it had come up tails for George Low in Apollo 8, people would have said, "What on Earth are you doing trying to send the second manned flight of an Apollo spacecraft around the Moon?" If the first flight of the Saturn V on the all-up had failed, people would have said, "Well, that was really dumb to try to test all three stages at once." The first time it had ever been done, everybody told him he should not do it, and look what happened.

The Apollo Program was audacious, and occasionally it failed. But the only reason we had a spacecraft as mature as the one we had in 1967 was because Joe Shea had been operating that way for four years and accomplishing wonderful things by so doing.

In trying to pull together my thoughts about the way NASA operated, I would like to suggest considering the Apollo 12 mission. I recommend that NASA have a three-day seminar for senior management staff on Apollo 12, meditating on it as a fascinating example of managing a space program. As some may recall, Apollo 12 was hit by lightning. It was actually hit by lightning twice in the boost phase of the first stage, knocking everything onboard to flinders. All the warning lights went on. Down on the ground, it wasn't that all the data had been lost on the controllers' screens, but that the data made no sense whatsoever. They didn't know that the spacecraft had been hit by lightning. All the controllers knew was that the platform had been lost; the guidance platform had been lost; that they weren't able to read any of their data; and it was taken for granted that what you had to do at that point was abort.

Shortly after liftoff on 14 November 1969, lightning struck the Apollo 12 Saturn V launch vehicle and the launch tower. NASA Image KSC-69PC-812. Special thanks to Kipp Teague for help with this image.

Here is the first vignette from that Apollo 12 launch. Sitting at one of the mission control consoles was one John Aaron. He later rose to great heights in NASA, but at that time he was only twenty-five or twenty-six years old. A year earlier, he had been sitting in the control room at Houston watching a test at the Cape, which they often did just to get to understand their systems better.

This particular time, at some point during the test, his screen suddenly turned to weird numbers. Incidentally, the screens of the Apollo controllers did not have nice graphics on them. They were black screens. They had fuzzy white numbers, [with] columns of fuzzy white numbers on them at that time. That's all the controllers viewed. The numbers were constantly changing. Incidentally, it is also true that the numbers were not in real time because the computers were quite slow. So controllers had to factor in that some of the numbers that were changing were 15 seconds old,

while other numbers were 10 seconds old and so forth. That is the kind of thing you did if you were an Apollo controller.

Aaron had looked into things, called up the Cape, and finally managed to figure out what was going on. He was told of an obscure board, called the signal conditioning equipment, SCE, that would have restored their numbers if it was switched to auxiliary mode. This was something that John Aaron had

Technicians in the Firing Room listen to Apollo 12 and Mission Control overcome lightning-induced electrical problems. NASA Image KSC-69P-856. Special thanks to Kipp Teague for help with this image.

done that was not a formal part of his job. It was part of hundreds of similar experiences he'd had. This was not something that the controllers had practiced in any simulation since then. He was probably one of the only people in all of NASA who knew this thing existed. In the critical launch phase, when they were about to lose a crew, when everything was going crazy, Aaron looked at that screen, and he understood within a matter of seconds what was going on.

On the Apollo 12, the spacecraft had been hit by lightning twice in the initial ascent phase. Controllers had lost the platform but managed to reset it. They had a couple of hours in which to go through tests of the spacecraft, and then they had to decide whether to go forward with translunar injection.

Catherine Cox and I really wanted to reconstruct the decision that was being made by Rocco Petrone, Chris Kraft, Jim McDivitt, and the other senior people who were in charge of that flight. We talked to all of them, and we couldn't get a story out of it because here's what happened. These twenty-six-, twenty-seven-, twenty-eight-, twenty-nine-year-old controllers went through all the systems down there in the control room. Then they turned around to the back row and said, "We've got a clean spacecraft; let's go," and there was no fretting about it.

When I was interviewing Gene Kranz once, I asked him, "Gee, this seems to me like a very dicey thing to do. Yes, you've checked out the spacecraft, but, after all, the thing has been hit by a huge bolt of lightning through all its electronics." Kranz was very matter-of-fact about it—"No, you go the way the data leads you." So I finally asked him "if a similar thing happened with the Space Shuttle and you had to make the equivalent of a decision to go out of Earth orbit, would you do the same thing?" Gene Kranz was not often at a loss for an answer, but he just sat and stared at me for about five seconds, and then he broke into a laugh, and he didn't say anything.

That was the way that that mission worked. It was a story of everything that made the human spaceflight program such a wonderful adventure, as well as an excellent case study from which later people could learn.

I second the remarks of Administrator Goldin about the future of human spaceflight. I think that his aspirations for it are just right. The only thing I would add is that if it is to succeed, human spaceflight must most of all capture the public imagination.

Part of the reason for that is hardheaded politics; you can't have a big program unless you have gotten the political funding for it, and the political funding only comes for it if you have captured a large part of the public imagination.

The essence of human spaceflight is that it does great things. That is how it captures the public imagination. About 600 years ago, with the invention of the scientific method, the deep abiding human impulse to understand and to explore, which previously had been confined to philosophy and religion, was let loose on all the other ways that we could explore the world. Now, in the twentieth century, I think that human space-flight touches the wellspring of the human spirit and excites a great many people. Human spaceflight also represents the great next adventure in that continuing quest to understand and to explore—only this time it is to understand and explore the universe.

We are never going to get a majority of the American people to share in that aspiration any more than you could get 51 percent of the people in Europe who wanted to get in small dangerous boats and go to the new world. There always will be objections such as "We would be better off spending money to combat poverty here on Earth." There is, however, a sizable minority who has a lot of influence, and they can be energized. But the only way that they can be energized is if human spaceflight remains true to its mission—it must do something beyond building one brick after another. It must continue to push the envelope with audacity, by going [to] new places, by doing new things, by taking on grand missions.

So as somebody who doesn't have a technical background and doesn't work for NASA, I'll go ahead and give some advice anyway. Get a grand mission, believe in it, give it to a new generation, and get the hell out of the way. Thank you very much.

*The Spaceflight Revolution
Revisited—William Sims Bainbridge*

Hermann Oberth in the foreground appears with officials of the Army Ballistic Missile Agency at Huntsville, Alabama, in 1956. Left to right: Dr. Ernst Stuhlinger (seated); Major General H.N. Toftoy, Commanding Officer for Project Paperclip; Dr. Wernher von Braun; and Dr. Robert Lusser. NASA Image CC-417.

There are two models of the future of spaceflight, and there are two theories of how that future might be achieved. The first *model* of spaceflight assumes that we have already achieved most of what is worth achieving in space, whereas the second imagines it will be possible to build a truly interplanetary civilization in which most human beings live elsewhere than on Earth. The first *theory* holds that progress comes incrementally from the inexorable working of free markets and political systems, whereas the second believes that revolutionary transformations must sometimes be accomplished by social movements that transcend the ordinary institutions and motivations of mundane existence.

My 1975 Harvard doctoral dissertation, published in 1976 as *The Spaceflight Revolution*, attributed the early stages of development of space technology in large measure to a social movement that transcended ordinary commercial, military, or scientific motives.[1] First, visionaries like Konstantin Tsiolkovsky, Robert Goddard, and Hermann Oberth developed the ideology of spaceflight. Then tiny volunteer groups coalesced around their ideas in Germany, America, Russia, and Britain, becoming the vanguard of a radical social movement aimed at promoting the goal of interplanetary exploration. Shrewd and dynamic entrepreneurs, notably Wernher von Braun and Sergei Korolev, took the movement on a military detour, gaining the support of

1. William Sims Bainbridge, *The Spaceflight Revolution* (New York: Wiley Interscience, 1976).

the German and Russian governments. Finally, the movement became institutionalized as the space programs of the Soviet Union, United States, and other countries.

After I wrote, some historians gave greater emphasis to the technical needs of the German war machine and the technocratic values of the Soviet Union in the development of spaceflight.[2] Their analyses focus on later phrases in space history, and certainly the social movement was crucial at the very beginning. There is room to debate how long it was influential and when institutional factors took control. The role of a transcendent social movement in the development of spaceflight is an intrinsically interesting question for historians, but it becomes very important if we use the past to try to understand the future. Thus, for me, the crucial question has always been "Can spaceflight technology develop to the fullest possible extent without the often irrational impetus that a social movement can contribute?"

Human beings have not left low-Earth orbit since 1972, and for thirty years the emphasis in space has been relatively modest projects that satisfy some of the conventional needs of terrestrial society. The 1986 report of the National Commission on Space argued that the solar system is the future home of humanity, where free societies will be created on new worlds, and great new resources will benefit humanity.[3] However, governments, private

2. Michael J. Neufeld, *The Rocket and the Reich: Peenemünde and the Coming of the Ballistic Missile Era* (New York: Free Press, 1995); Walter A. McDougall, *The Heavens and the Earth: A Political History of the Space Age* (New York: Basic Books, 1985).
3. National Commission on Space, *Pioneering the Space Frontier* (New York: Bantam Books, 1986).

enterprise, and the general public have not endorsed solar system colonization as a practical or worthy goal.

This essay will first consider whether technological breakthroughs in space technology and the rational motives of ordinary institutions have the capacity to break out of this relatively static situation. Then we will survey the roles that social movements of various kinds might play and conclude with an examination of one particular nascent movement that might possibly build the foundation for a spacefaring civilization.

When *The Spaceflight Revolution* was written, we had great hopes that the Space Shuttle would be an economic as well as technical success, but sadly, the cost of launching to Earth orbit remains prohibitively high for many applications. The most recent disappointment is the cancellation of the X-33 and the inescapable realization that we are still a long way from the ability to develop a low-cost launch system.[4]

Science-fiction writers and other visionaries have suggested a vast array of alternative orbital launch methods.[5] Some, like electric catapults and Jacob's ladders, have some grounding in scientific principles but may present insurmountable engineering difficulties. Others, like antigravity and reactionless drives, have no basis in science and thus must be presumed impossible. A third of a century ago, practical nuclear fission rockets were

4. Frank Morring, Jr., "NASA Kills X-33, X-34, Trims Space Station," *Aviation Week and Space Technology* (5 March 2001), pp. 24–25.
5. William Sims Bainbridge, *Dimensions of Science Fiction* (Cambridge: Harvard University Press, 1986).

under development, but this approach now seems environmentally unacceptable. It is hard to devise a more environmentally benign propellant than the hydrogen and oxygen used by the main engines of the Space Shuttle.

There is some hope that nanotechnology will save the day with materials based on carbon nanotubes that are vastly stronger yet lighter than metals.[6] However, the X-33 failure shows that it is not easy to work with radically new structural materials in demanding aerospace applications, and we may be many decades away from being able to manufacture propellant tanks, wings, and other large structures from carbon nanotubes.

Perhaps Robert Zubrin is right that [the] use of native Martian resources will significantly reduce the cost of a manned expedition.[7] However, the cost may still be more than people are willing to invest. Thus, the Mars society that has been organized around Zubrin's vision may be more important for reviving the spirit of the spaceflight movement than for any particular technical innovation it offers.

Technological breakthroughs in rocketry would certainly help promote space development, but the advances we are likely to see over the next several decades will not be sufficient in themselves. We also need a profound boost in the motivation to invest in an aggressive space program.

6. Richard W. Siegel, Evelyn Hu, and M. C. Roco, *Nanostructure Science and Technology* (Dordrecht, Netherlands: Kluwer, 1999); M. C. Roco, R. S. Williams, and P. Alivisatos, *Nanotechnology Research Directions* (Dordrecht, Netherlands: Kluwer, 2000); M. C. Roco and William Sims Bainbridge, *Societal Implications of Nanoscience and Nanotechnology* (Dordrecht, Netherlands: Kluwer, 2001).
7. Robert Zubrin and Richard Wagner, *The Case for Mars* (New York: Free Press, 1996).

Satellites in low-Earth and synchronous orbit are of great importance in the collection and distribution of information, thus essential to the information economy. The wide range of civilian applications includes telephone, data transmission, television, navigation, weather observation, agriculture monitoring, and prospecting for natural resources.[8] The technology is largely perfected, and incremental progress can be achieved by improvement in information systems and simply by investing in more relatively small satellites of the kinds we already have.

Current space technology has proven the capacity to send robot space probes to any location in the solar system and a few billion miles beyond. Orbiting observatories, such as the decade-old Hubble Space Telescope, are effective means for gaining information about the vast realm that lies beyond the reach of space probes.[9] Much can be accomplished over the next century in space science without the need for major new launch technology. Indeed, one could argue that if science were the prime purpose of spaceflight, we would have done well to keep manufacturing the forty-year-old Saturn I, rather than developing more sophisticated launch systems.

Many scientists and ordinary citizens believe that the chief justification for the space program is the knowledge of our place in the universe gained by probes and space telescopes. However,

8. For example, see the National Research Council report, *People and Pixels: Linking Remote Sensing and Social Science*, ed. Dianna Liverman, Emilio F. Moran, Ronald R. Rindfuss, and Paul C. Stern (Washington, DC: National Academy Press, 1998).
9. Robert W. Smith, *The Space Telescope* (New York: Cambridge University Press, 1989).

if the government really wanted to advance fundamental knowledge that is interesting to the general public as well as to scientists, it would put its money not into spaceflight but into paleontology, archaeology, and anthropology—extremely underfunded fields where rapid advances could be expected to follow quickly from any increased investment.

The search for human origins is a noble and tremendously exciting scientific initiative waiting for the political will to achieve profound discoveries. Very little is currently invested in primary data collection in paleontology and archaeology, and a few million dollars a year could work wonders. In physical anthropology, tools of genetic science already exist that could chart the evolution of the human species and its geographic dispersion. For example, existing techniques are capable of sequencing the DNA of Neanderthal specimens and determining their relationship to modern humans.[10] All that is needed is funding.

Military reconnaissance satellites have been essentially perfected, and they are already capable of accomplishing almost any data gathering the defense establishment is willing to invest in.[11] For a quarter century, enthusiasts have urged the development of a space-based missile defense system, perhaps employing beam weapons. If it required orbiting many large installations, it

10. Matthias Krings, Anne Stone, Ralf W. Schmitz, Heike Krainitzki, Mark Stoneking, and Svante Pääbo, "Neandertal DNA Sequences and the Origin of Modern Humans," *Cell* (1997) 90: 19–30; see also Dennis H. O'Rourke, S. W. Carlyle, and R. L. Parr, "Ancient DNA: Methods, Progress, and Perspectives," *American Journal of Human Biology*, 1996, 8: 557–571.
11. Jeffrey T. Richelson, *America's Space Sentinels: DSP Satellites and National Security* (Lawrence: University of Kansas Press, 1999).

would promote the development of efficient launch vehicles which could then be applied to other purposes.[12] But currently its advocates emphasize localized theater defense systems and "smart rock" ICBM interception methods that do nothing to advance civilian spaceflight.

Since the 1960s, there has been much talk about commercial exploitation of outer space. For a time, attention was given to the idea of collecting solar energy in space and beaming it to Earth, and there still is hope that some new industrial processes that require weightlessness will prove to be economically profitable. However, space-based solar energy systems would be extremely costly and are not currently part of the world's response to energy needs.[13] Today, materials scientists are much more excited about a wide range of terrestrial nanotechnology techniques than about the dubious value of weightless manufacturing.

Like military applications, hypothetical industrial satellites would probably be in low-Earth orbit; although, some writers have argued that it might be cheaper to build them from lunar materials because of the low velocity required to leave the Moon.[14] This would demand some degree of lunar colonization, and it would thereby build a transportation infrastructure that would reduce the cost of deep-space missions.

Nonetheless, it is very difficult to develop a scenario in which the Earth itself could ever benefit from importation of raw materials

12. William Sims Bainbridge, *The Spaceflight Revolution* (New York: Wiley Interscience, 1976), pp. 241–243.
13. *Electric Power from Orbit: A Critique of a Satellite Power System* (Washington, DC: National Academy of Sciences, 1981).
14. Gerard K. O'Neill, *The High Frontier* (New York: Bantam Books, 1977); Richard D. Johnson and Charles Holbrow, editors, *Space Settlements: A Design Study* (Washington, DC: National Aeronautics and Space Administration, 1977).

from beyond the Moon. It is more than a cliché that the world is becoming an information society, postindustrial rather than industrial.[15] The Earth has ample supplies of almost every useful chemical element, and it is not plausible that we could find energy sources on Mars that would be cost-effective to bring to Earth. Martian resources would be of value if we had already decided to live there, but we would need some motivation other than raw materials to do so. In purely economic terms, beyond synchronous orbit or maybe lunar orbit there may be no bucks; therefore, no Buck Rogers.

Some say that the pressure of population growth on Earth will force humanity to colonize other worlds. Perhaps the most plausible version of this scenario was suggested in Kim Stanley Robinson's series of novels about terraforming Mars—the rich ruling classes might want to develop Mars as a home for themselves when Earth becomes unendurably overpopulated.[16]

Unfortunately, examination of actual fertility and mortality trends does not provide a clear demographic justification for space colonization. The population explosion has not yet halted in many poor nations, but they certainly do not have the wealth for spaceflight. Fertility rates have dropped so far in most of the industrial nations that they are poised for a population collapse that would remove their motivation to expand out into space.

Recent United Nations estimates predict that nineteen nations of the world will each lose more than a million in population

---

15. Daniel Bell, *The Coming of Post-Industrial Society* (New York: Basic Books, 1973).
16. Kim Stanley Robinson, *Red Mars* (New York: Bantam, 1993); *Green Mars* (New York: Bantam, 1995); *Blue Mars* (New York: Bantam, 1997).

by the year 2050: Russia (loss of forty-one million people), Ukraine (twenty million), Japan (eighteen), Italy (fifteen), Germany (eleven), Spain (nine), Poland (five), Romania (four), Bulgaria (three), Hungary (two), Georgia (two), Belarus (two), Czech Republic (two), Austria (two), Greece (two), Switzerland (two), Yugoslavia (two), Sweden (one), and Portugal (one).[17] Fertility rates are also already below the replacement level in Australia, Canada, France, New Zealand, and the United Kingdom.

Fertility rates are still above replacement in the United States, and the U.S. Census projects population growth throughout the next century.[18] In part, growth is assured by immigration and by the fact that fertility rates are still high in some minority groups. Major uncertainties are the roles of religion and politics. The collapse in European fertility rates may partly be explained by secularization and by indirect effects of the welfare state.[19] America is far more religious than almost any European nation today, other than Ireland and Poland, and its political environment is quite different from that of Europe or Japan. If the United States eventually follows the other industrial nations in abandoning religion and adopting the welfare state, then American fertility rates could collapse just as those in most of

---

17. Population Division, Department of Economic and Social Affairs, United Nations, *World Population Prospects: The 2000 Revision* (New York: United Nations, 2001), p. 58.
18. Frederick W. Hollmann, Tammany J. Mulder, and Jeffrey E. Kallan, "Methodology and Assumptions for the Population Projections of the United States: 1999 to 2100," Population Division Working Paper No. 38, United States Census Bureau, 2000.
19. Ronald Inglehart and Wayne E. Baker, "Modernization, Cultural Change, and the Persistence of Traditional Values," *American Sociological Review* (February 2000), 65: 19–51.

Europe have already done. If that happens, then there is no nation both rich enough and demographically motivated to colonize the solar system.

Finally, one might hope simply that the passage of time will allow a steadily increasing portion of the population to become interested in space. Spaceflight accomplishes a little more each year, and the growing status of science fiction in popular culture should also contribute to increased enthusiasm.

However, opinion polls reveal only modest growth in support for the space program. Perhaps the best data source is the General Social Survey, a repeated scientific study of a random sample of Americans that has included a question about the space program for twenty-five years. In 1973, just 7.8 percent of the American public wanted funding for the space program increased. By 1998, this fraction had grown just to 10.8 percent. A pessimistic way to look at this is to note that this increase of 3 percentage points over a quarter century would mean 12 percentage points every century. Linear extrapolation would predict a majority of the population would support increased space funding in about the year 2325.

Of course, a crude projection like that is scientifically indefensible. Support has moved up and down over the years, apparently in response to events. The highest level of support was in 1988, responding to the nation's return to space after the *Challenger* disaster, when 18.9 percent wanted funding increased. The biggest trend over the twenty-five years was actually a shift from feeling funding should be reduced to feeling it was about right. In 1973, 61.4 percent wanted the space program reduced, compared with only 42.2 percent in 1998. Those who felt about the right amount was being invested rose from 30.8 percent in 1973 to 43.8 percent in 1998. But a projection based on the fifteen years from 1983 to 1998 shows no growth in those who want space funding increased and no decline in the proportion of the population who want it reduced; so projections are very sensitive to the assumptions on which they are based.

While opinion polls give some reason for slight optimism, they certainly do not reveal the kind of rapid growth in support that would be required to break out of the current doldrums. Hope springs eternal, but there is little reason to expect that either a breakthrough in space technology or a surge in conventional motivation will transform spaceflight in our lifetimes. Thus, we need to consider the possible impact of another spaceflight social movement.

The regularities of human interaction can be classified in terms of four levels of social coordination—*parallel behavior*, collective behavior, social movements, and societal institutions.[20] Parallel behavior is when individuals do roughly the same thing for similar reasons, but without influencing each other directly. An example is the isolated pioneers who developed the intellectual basis of spaceflight, including Tsiolkovsky, Goddard, and the early work of Oberth. On the basis of their ideas, an international network of informal communication developed, chiefly

20. William Sims Bainbridge, "Collective Behavior and Social Movements," in *Sociology* by Rodney Stark (Belmont, California: Wadsworth, 1985), pp. 492–523, reprinted in second edition (1987) and third edition (1989).

through publications, in which the ideas were disseminated, and spaceflight enthusiasts came into contact with others of like mind. The sociological term for informally coordinated mass activity is *collective behavior*, including such phenomena as panics, riots, fads, and crazes.

It often happens that collective behavior can develop a degree of formal organization and become a *social movement*. For spaceflight, the watershed was the founding of prospace voluntary organizations, notably in Germany, the United States, the Soviet Union, and Britain. A successful social movement often becomes incorporated in or co-opted by a *societal institution*, such as government space programs. Then, the early enthusiasms of the typical institutionalized movement become mired in bureaucratic inertia, and it is very difficult to transform well-established institutions.[21]

Much of the traditional social-scientific literature on social movements focuses on the movements of deprived groups within society.[22] These often take the form of protests, and they typically challenge the comfortable status of societal elites. To many influential people, the evolutionary processes of conventional societal institutions feel safer and more reasonable than revolutionary movements.

Since the end of the Apollo program, a number of moderate social movement organizations have supported increased efforts

21. William Sims Bainbridge, "Beyond Bureaucratic Policy: The Space Flight Movement," pp. 153–163 in *People in Space*, ed. James Everett Katz (New Brunswick, New Jersey: Transaction, 1985).
22. Neil J. Smelser, *Theory of Collective Behavior* (New York: Free Press, 1962); Hans Toch, *The Social Psychology of Social Movements* (Indianapolis: Bobbs-Merrill, 1965); Ted Robert Gurr, *Why Men Rebel* (Princeton: Princeton University Press, 1970).

in space.[23] In the main, these are respectable groups, and their contributions have been worthwhile. However, as Michael Michaud noted in his study of these groups, they have not achieved significant breakthroughs.[24]

A really new spaceflight movement might upset the delicate relationship between the established space program and the branches of government that provide the money for it, and it might alienate many opinion leaders in the general public, even if it energized the enthusiasm of others. At the very least, a fresh social movement would demand fresh thinking that shatters conventional notions about what science, technology, and the human spirit could accomplish in space.

Religious movements are especially suspect in the modern era, yet they have the capacity to break through ordinary routines and to experiment with utopian alternatives such as [an] extraterrestrial society.[25] Few people already involved in the space program, and few members of the general public, are prepared to embrace a radically new religion. Some of them are faithful believers in the traditional religions. Most of the rest are probably secularists with neither religious faith nor much trust in religious enthusiasts.

Most people seem horrified by the few highly publicized religions oriented toward contact with extraterrestrial beings.

23. Trudy E. Bell, "American Space-Interest Groups," *Star and Sky* (September 1980), pp. 53–60.
24. Michael A. G. Michaud, *Reaching for the High Frontier: The American Pro-Space Movement*, 1972–84 (New York: Praeger, 1986), p. 308.
25. William Sims Bainbridge, "Religions for a Galactic Civilization," pp. 187–201 in *Science Fiction and Space Futures*, ed. Eugene M. Emme (San Diego: American Astronautical Society, 1982).

Both Heaven's Gate[26] and The Solar Temple[27] tried to travel to other worlds by committing suicide, and the latter also committed a number of murders. A theologically similar space-oriented group called the Raelian Movement has not resorted to violence but has hurled a powerful religious challenge at conventional society by setting out to clone human beings as part of its radical method for transcending the limitations of terrestrial life.[28]

Religious movements have a tendency to pursue goals by supernatural rather than natural means. An example is the little book published by the Hare Krishna movement, *Easy Journey to Other Planets*, advocating chanting rather than rocketry as the best means to experience other worlds.[29] Thus, it is possible that space-oriented cults will seek to explore the galaxy, but they will

26. Robert W. Balch, "When the Light Goes Out, Darkness Comes," in *Religious Movements*, ed. Rodney Stark (New York: Paragon House, 1985), pp. 11–63; "Waiting for the Ships: Disillusionment and the Revitalization of Faith in Bo and Peep's UFO Cult," *The Gods Have Landed: New Religions From Other Worlds,* ed. James R. Lewis (Albany: State University of New York Press, 1995), pp. 137–166; Ryan J. Cook, "Heaven's Gate," *Encyclopedia of Millennialism and Millennial Movements*, ed. Richard Landes (New York: Routledge, 2000), pp. 177–179; Winston Davis, "Heaven's Gate: A Study of Religious Obedience," *Nova Religio* 3 (2000) *http://www.novareligio.com/ davis.html*
27. Susan J. Palmer, "Purity and Danger in the Solar Temple," *Journal of Contemporary Religion* 11 (1996), pp. 303–318; "The Solar Temple," *Encyclopedia of Millennialism and Millennial Movements,* ed. Richard Landes (New York: Routledge, 2000), pp. 394–398.
28. Susan J. Palmer, "The Raëlian Movement International," *New Religions and the New Europe*, ed. Robert Towler (Aarhus, Denmark: Aarhus University Press, 1995), pp. 194–210; Phillip Charles Lucas, "Raelians" in *Encyclopedia of Millennialism and Millennial Movements*, ed. Richard Landes (New York: Routledge, 2000), pp. 342–344.
29. Bhaktivedanta Swami, A. C., *Easy Journey to Other Planets* (Boston: ISKCON Press, 1970).

probably attempt to do so through supernatural rituals rather than through spaceflight.[30]

This brief survey of research on social and religious movements is not very encouraging. However, the examples of the civil rights, women's liberation, and environmentalist movements remind us that social movements are often very effective in changing society's priorities. Perhaps a totally new kind of movement could emerge in the next few years, employing technology to serve fundamental human needs that in earlier centuries would have motivated religious or political movements.

Let us imagine a successful social movement of the future that could actually build an interplanetary and even interstellar civilization. I will present one idea here, but perhaps others are possible. The idea relies upon plausible developments in fields of science and technology that seem remote from astronautics—namely cognitive neuroscience, genetic engineering, nanotechnology, and information systems. But the fundamental key is a transcendental movement that would provide the motivation to apply these developments to the foundation of cosmic civilization.

The chief impediment to rapid development of spaceflight is the problem of returning a profit to the people who must invest in it. The most obvious way to motivate people to invest in interstellar exploration is to invite them to travel personally to

---

30. Rodney Stark and William Sims Bainbridge, *The Future of Religion* (Berkeley: University of California Press, 1985); *A Theory of Religion* (New Brunswick, New Jersey: Rutgers University Press, 1996); William Sims Bainbridge, *The Sociology of Religious Movements* (New York: Routledge, 1997).

the stars and create new lives for themselves on distant worlds. But we are decades and perhaps centuries away from having the technological capability and infrastructural base to accomplish this in the conventional manner we have always imagined—by flying living human bodies and all the necessities of life to other planets. There is, however, another possible way.

Visionaries in a number of cutting-edge disciplines have begun to develop the diverse toolkit of technologies that will be required to overcome death. A prominent example is Ray Kurzweil, a pioneer of computer speech recognition, who argues that human beings will gradually merge with their computers over the next century, thereby becoming immortal.[31] The idea dates back at least to Arthur C. Clarke's 1953 novel, *The City and the Stars*. In 1966, Roger MacGowan and Frederick Ordway speculated that successful spacefaring species might evolve past the state of being biological organisms, becoming "intelligent synthetic automata."[32] We have in fact advanced some distance in that direction over the past thirty-five years, and we now see the real possibility of achieving that dream in a manner that preserves unique human personalities and blends natural with synthetic modalities.

For a number of years, I have studied the techniques for archiving aspects of human personality in computerized infor-

31. Ray Kurzweil, *The Age of Spiritual Machines: When Computers Exceed Human Intelligence* (New York: Viking, 1999).
32. Roger A. MacGowan and Frederick I. Ordway, III, *Intelligence in the Universe* (Englewood Cliffs, New Jersey: Prentice-Hall, 1966).

mation systems, along the way publishing computer-assisted textbooks on some of the methodologies.[33] In May 1997, I launched a Web-based project, called The Question Factory, to create a very large number of questionnaire measures to archive aspects of personality that were generally missed by standard psychological tests.[34] In addition to placing a number of item-generation open-ended surveys on my own Web site, I joined the team creating *Survey2000* and *Survey2001*, two major online questionnaire projects sponsored by the National Geographic Society. My initial result was a set of eight personality-archiving software modules incorporating 15,600 items and 31,200 measurements.[35] Anyone can begin to archive his or her personality using these Windows-based programs today.

A complementary approach involves making digital audio-visual recordings of a person's perceptions, speech, and behavior. For example, Carnegie-Mellon University's Experience on Demand project is developing "tools, techniques, and systems allowing people to capture a record of their experiences unobtrusively."[36] Steven Spielberg's Survivors of the Shoah Visual History Foundation has videotaped the reminiscences of more than 52,000 survivors of the European holocaust, a 180-terabyte

33. William Sims Bainbridge, *Survey Research: A Computer-Assisted Introduction* (Belmont, California: Wadsworth, 1989); this textbook includes nine software programs and datasets; *Social Research Methods and Statistics* (Belmont, California: Wadsworth, 1992); this textbook includes eleven software programs and datasets.
34. William Sims Bainbridge, "Religious Ethnography on the World Wide Web," *Religion on the Internet*, ed. Jeffrey K. Hadden and Douglas W. Cowan, Vol. 8 of the annual *Religion and the Social Order* (New York: JAI/Elsevier, 2000), pp. 55–80, especially pp. 66–75.
35. The Question Factory, *www.erols.com/bainbri/qf.htm*
36. *www.informedia.cs.cmu.edu/eod/*

dataset that cost $175 million to assemble.[37] The same effort could have captured much of the personality of a single individual. A combination of real-time computer graphics and artificial intelligence based on an individual's full personality record could even today produce a realistic dynamic simulation of that individual.

Many people today carry personal digital assistants (PDAs), and some of these are already connected to Internet. Over the next few years, these will evolve into multimedia gateways to the world of information, serving as advisors, coaches, agents, brokers, guides, and all-purpose servants. At the same time they perform all these functions, they can unobtrusively record the user's wishes, thoughts, actions, and words. Advanced devices of this type will adapt to the user's needs and personality, so they will have to learn many of the facets of the person anyway. They will also be companions that converse and play games with the user. Many forms of personality-archiving methods can be blended seamlessly with these activities.

A combination of foreseeable advances in several fields of science and technology will permit vast improvements in our ability to capture and reanimate a human personality. In time, cognitive neuroscience, perhaps drawing upon molecule-size sensor developments in nanotechnology, will be able to chart the structure and function of a living human brain. "Gene on a chip"

37. *www.vhf.org/*

bioelectronic devices will permit cost-effective sequencing and analysis of those aspects of a person's genetic code that influence his or her personality. Information science, especially in the very active field of digital libraries, will develop the necessary techniques for efficient storage and access of petabyte records of the individual. Finally, advances in genetic engineering, information systems, and robotics will allow archived human beings to live again, even in transformed bodies suitable for life on other planets and moons of the solar system.

New lives must be lived on new worlds.[38] Overpopulation from a zero death rate would soon fill any one planet, and humanity would lose its finest treasure if there were no more children. In the past, several religions imagined that the afterlife was located in Heaven. Once reanimation of archived human personalities becomes possible, it will be necessary to enact a worldwide constitutional law that resurrection must not be done on Earth, but only in the heavens.

We see the beginnings of this prohibition against terrestrial resurrection in the remarkably powerful worldwide movement to ban human reproductive cloning. Other technologies are likely to be banned on Earth in later decades, such as advanced forms of artificial intelligence and android robots. Genetic engineering is already under concerted attack, and there are the

38. William Sims Bainbridge, *Goals in Space* (Albany, New York: State University of New York Press, 1991).

beginnings of a movement to ban some forms of nanotechnology.[39] Scientists in these fields may have to do their work beyond the reach of terrestrial religions and governments, but that will be costly. Only a goal as valuable as eternal life could motivate investment in substantial scientific infrastructure on the Moon or Mars.

Calculation of the geometric realities facing colonization of the universe suggests that there might not be enough room in the galaxy for endless copies of absolutely everybody. The population of an expanding sphere of inhabited worlds increases according to the cube of its radius, while the surface area from which colonization ships can directly reach new solar systems increases only as the square of the radius.[40] To some extent, this problem can be dealt with by gradually increasing the time between lives. But unless a means of instantaneous interstellar travel is devised, the rate of expansion of the human population is limited.[41]

The answer is a simple one. A person must earn a new life by contributing in some way, direct or indirect, to the development and maintenance of the entire system that explores and colonizes space. Thus, each generation has a moral contract with the ones that follow. Every person who contributes substantially has a right to expect at least one more life. Future generations must honor that promise if they are to have any hope that the generations after them will grant them a second life as well.

39. Bill Joy, "Why the Future Doesn't Need Us," *Wired* (April 2000).
40. William Sims Bainbridge, "Computer Simulation of Cultural Drift: Limitations on Interstellar Colonization," *Journal of the British Interplanetary Society* (1984) 37: 420–429.
41. Sebastian von Hoerner, "Population Explosion and Interstellar Expansion," *Journal of the British Interplanetary Society* (1975) 28: 691–712.

As noted above, many human populations are failing to reproduce even at the replacement level and are destined to vanish gradually from the Earth through an insidious form of genetic suicide.[42] In particular, highly educated nations and groups whose religion or philosophy does not encourage childbirth are failing, whereas uneducated populations and fundamentalist groups are growing. Well-educated people can ensure the demographic growth of their population through interstellar immortality. By "arrival of the fittest," those with the most advanced minds and cultures will spread across the galaxy.

Even a very low birthrate per lifetime can cause population growth when an individual has many lifetimes in which to reproduce. Additionally, some individuals who make extraordinary contributions to human progress may thereby earn the right to live out several lives simultaneously in different solar systems, reproducing themselves as well as giving birth to children who are distinct personalities.

We have the technology, already today, to begin archiving human personalities at low fidelity within what I call Starbase, a database destined eventually to be transported to the stars. To gain entry to Starbase, a person must contribute significantly in some way to the creation of interstellar civilization. One way is to help develop technologies for archiving and reanimating

---

42. Nathan Keyfitz, "The Family that Does not Reproduce Itself," *Below Replacement Fertility in Industrial Societies*, ed. Kingsley Davis, Mikhail Bernstam, and Rita Ricardo-Campbell (Cambridge, England: Cambridge University Press, 1987), pp. 139–154; Ben J. Wattenberg, *The Birth Dearth* (New York: Ballantine Books, 1987).

human personalities at ever higher fidelity. Another is to work toward the establishment of small human colonies, first on the Moon and Mars, where Starbase can be headquartered and where serious work on reanimation can begin.

When the time comes for the first interstellar expeditions, they will be carried out not by biologically based humans in their first brief lifetimes, but by eternal Starbase modules incorporating the archived but active personalities of the crew and colonists. At the destination, the crew will not waste its time terraforming planets, but will adapt the colonist into whatever form (biological, robot, cyborg) [that] can thrive in the alien environment. Subsequent waves of colonists can be sent as radioed data files in a technically feasible version of the old science-fiction dream of teleportation.

A Starbase movement could offer the stars to people living today, and this realistic hope would motivate us to create first an interplanetary then an interstellar civilization. It draws upon advanced technology from fields other than rocketry, and it promises to serve the instinctive desire for survival. By conceptualizing human beings as dynamic systems of information, it harmonizes with the fundamental principles of postindustrial society. Such a movement could provide powerful new motivations for a second spaceflight revolution.

In conclusion, ancient Greek scientists knew that the Earth was a sphere, and they understood roughly how large it is. However, the classical civilization of Greece and Rome failed to exploit that knowledge, send expeditions to the Americas, and colonize the New World. Similarly, our more technically advanced civilization understands the fundamental scope of the galaxy, yet we seem to lack the cultural dynamics and social organization required for interplanetary let alone interstellar travel and settlement.

Pessimists might conclude that we should tear down our present civilization quickly to hasten the next Dark Age, so that the successor spacefaring civilization will get an earlier start. But the seeds of each new civilization need to be securely planted within the old—just as Christianity took root within classical society and later helped shape industrial society. Two thousand years ago, Christianity was but one of many cults vying for attention within the Roman Empire, but it rose to become the most influential movement of all human history. Thus, optimists would attempt to launch many space-related social movements in the hopes that one of them would eventually take humanity to the stars.

At the extreme, optimists and pessimists might agree that the human species, as it is currently defined, simply is inferior to the task. With a lifespan generally under a century, we require quick returns on our investments, and our instincts are too easily satisfied by modest lives on our home planet. But extreme optimists differ from pessimists in that they imagine we can evolve into something higher, a truly cosmic species for whom all the universe is home.

Count me among the optimists. Probably, many intellectual leaders and policymakers in the standard aerospace agencies and corporations will find the Starbase idea too radical for their tastes. Yet business as usual is not going to create interplanetary civilization. In time, the standard institutions of Western civilization will disintegrate, like those of the Roman Empire 1,600 years earlier. Already we see demographic trends that are extremely worrying—unchecked population growth in the poor countries and impending collapse in most advanced nations. Human exploration of the universe through an aggressive space program has nearly stalled. The future demands a new spaceflight social movement to get us moving again.

Mutual Influences: U.S.S.R.-U.S. Interactions During the Space Race—Asif Siddiqi

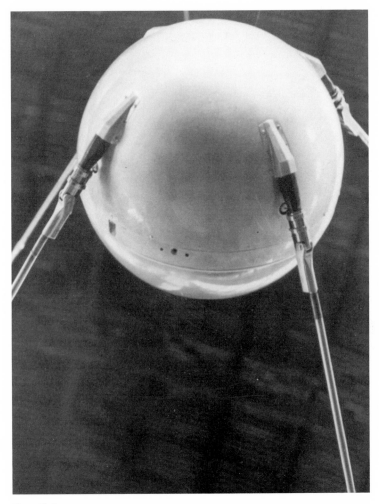

Sputnik 1.

I would like to take a broader historical view of the space race and look at the relationship between the Soviet Union and the United States in the early years of the space race. Then I would like to add some thoughts on the writing of history and how we understand it.

In the past ten years, our view of the space race has changed dramatically. Much of this has had to do with the fall of the Soviet Union and the subsequent availability of an unprecedented amount of information that has allowed us to rewrite that one side of the history of the space race. Previously, we only knew bits and pieces of what the Soviets did. Now we know not only what they did, but why they did certain things, which is an important aspect of writing history. Writing history is about making sense. It is about building patterns, about putting together pieces and making those pieces fit. It is not about chronologies. The writing of this new history indicates a fundamental maturity of our field and space history. We are now able to move from chronologies to making sense.

One of the things that I want to talk about today is how we have understood the space race. Traditionally, we have viewed it in terms of action and reaction. One side reacted to the other and did certain things, and then the other side reacted to that. So there was this chain reaction of events.

The new historical record suggests that's not so far from the truth, but perhaps we need a slightly more nuanced approach. I would like to touch on three very important milestones in the space race and reexamine those events in the light of new information—Sputnik, the flight of Yuri Gagarin in 1961, and the Moon race.

Sputnik has been considered the first big milestone in the space race. For over forty years now, we have considered Sputnik the first shot, the opening salvo. I would not disagree that Sputnik was the first physical manifestation of the space race, but I would argue that the space race actually began before Sputnik. As most of you know, Sputnik was launched during the International Geophysical Year, a period of intense scientific research organized by scientists all over the world. There were a number of key proposals from the American side to participate during the International Geophysical Year [IGY].

As most of you know, the Eisenhower Administration announced in July 1955 that the United States would launch a satellite during the IGY. The reasons behind that decision are fairly complex, and so I will not go into that.

But what's most interesting from the Soviet side is how they reacted to this announcement. This announcement by the Eisenhower Administration set up a series of deliberations on the Soviet side about how they should react. These deliberations culminated in a project to preempt the American side by launching a huge scientific observatory. So, for the Soviets, the race had already begun immediately after the Eisenhower Administration's announcement.

An interesting sidebar to this occurred in late 1956, when Wernher von Braun's team-tested a missile. The Soviets mistakenly believed that this missile was actually trying to launch a satellite, which shook them. This misperception fueled a Soviet sense of urgency that "we have to do this before the Americans." Thus, they dropped their plans to launch this huge scientific observatory and decided to launch a small metal ball, which they could quickly do. Of course, we know that small metal ball as Sputnik.

So this new information asks us to reconsider and reframe certain events that we know as the "Holy Grail" of history. In one sense, the space race might not have begun on 4 October 1957, but rather it perhaps began two years earlier. That's an important distinction that may lead us to think about these events in a sharply different way.

The second issue is Yuri Gagarin's flight in 1961. Certainly apart from Sputnik, no other event has been more important for both sides in the early years of the space age. For the Soviets, this was their high point, their peak. For the Americans, Gagarin's flight was important because it set off deliberations that led to the decision to go to the Moon. Again, this demonstrates an action-reaction dynamic.

The new information also suggests that the Soviets really were reacting to the Americans, or at least what they thought the Americans were doing. Gagarin's flight was planned almost as a reaction to Mercury, and the timing of his flight was, in many ways, a reaction to what von Braun and others were thinking in terms of when NASA would launch the first American in space.

A lot of it had to do with timing, but a lot of it was pure luck. It could have easily been Alan Shepard who was the first human in space. It turned out to be Yuri Gagarin. But there definitely was an action-reaction dynamic, and it's important to take that into account in looking at other events in the space race too.

Finally, I would like to go to the third issue, which is the Moon race. We know that the Soviets were in a race to the Moon with the United States, and they tried hard. Kennedy committed

NASA to a Moon landing in 1961. It was a national goal. But the Soviets hardly took notice. In 1961, they had just launched Yuri Gagarin and had no reason to feel threatened. It was only in 1964 that they really began to think seriously about a Moon landing. It was a national priority only in 1967, which was too little, too late.

The action-reaction dynamic also plays into the Moon race. One of the interesting things that I have discovered in my research is how American information seeped over to the other side and how the Soviets dealt with it. Apollo is an interesting case because repeatedly throughout the 1960s, the Soviets simply did not believe that the Americans would make it to the Moon by 1969. They really had this feeling, and you would see this in documents. "Well, yes, they've got this equipment ready and that equipment ready, but it would just be impossible for them to make the 1969 deadline." What really shook them up was the Apollo 8 mission in December 1968, because this impressed upon the Soviets the imminent reality of a human Moon landing. But again, by then, it was too little, too late.

I think what all of this indicates is that, in some sense, the seeds of the Soviet failure were actually laid much earlier in the sense of complacency that emerged after Gagarin's flight. In some ways, the Soviets believed that "we're the best already," and it was too late before they realized that the U.S. was committed to Apollo and, thus, was a real threat.

Another interesting point concerns the post-Apollo period. The Soviets handled their failure in an unsurprising way, given that they had hidden their effort in the first place. They responded to Apollo by saying, "Well, we weren't in the race at all," and for many years, this denial was accepted lore for most Western observers. Because of this Western notion that "Well, we were just racing ourselves," many critiques of Apollo emerged. Whether or not one thought Apollo was a good or a bad thing for the ultimate future of the American space program, the value of it as an international competition and a demonstration of supreme rivalry was called into question for many years. During the 1970s and 1980s, many critics were frustrated and disappointed that "we've spent so much money and effort to get to the Moon first, and yet, there was no race after all."

Of course, in the past ten years, we have understood more clearly that there was indeed a space race. We know it was hard-fought, and we know the Americans won. I think this is one example of how history itself is dynamic and changing, pointing out that nothing is fixed. I expect that how we remember the Moon race forty years from now will be quite different from how we remember it today.

We should not compartmentalize history into saying that it is restricted by artificial boundaries and we can only understand history by looking through these blinders. We need to broaden our perspective by looking at the other side and trying to understand the action-reaction interrelationship that was going on in the 1960s and 1970s.

I would like to end with some final thoughts on how we evaluate history. Professional and academic historians often want to write about events and people from some measure of dispassionate distance. We tend to evaluate space history through contexts such as geopolitics, the Cold War, the missile gap, and presidential administrations. But there is also something to be said about imbuing history with the essence of what makes people want to do certain extraordinary things. If we look at the flight of Yuri Gagarin or the flight of Alan Shepard, it is almost impossible to see these as events outside of the Cold War.

But I think it is also important to recognize how important the flight of Yuri Gagarin, for example, was simply in the course of human history. It was the first time that a human being had left the planet Earth. I think that historians should not be afraid of appealing to that sense of the human imagination—to step back from geopolitics and the Cold War to see an event from a much broader perspective. I hope historians can take up that challenge in the future.

Making Human Spaceflight as Safe as Possible—Frederick D. Gregory

NASA's safety priorities.

From the first days of the Mercury program to today's efforts aboard the International Space Station, human safety has always been the primary consideration for human spaceflight.

Looking backward, consider NASA's first attempts to reach space without human crews. Rockets tipped over, rockets exploded on lift off, rockets careened off course . . . it sure didn't look safe.

Before we could put a life at risk, the rockets had to be made safer. How? Mostly through the application of brute-force engineering—essentially the "Fly, Fix, Fly" approach.

This approach did eventually lead to safer rockets; however, to produce a spacecraft intended for routine human flight into space, NASA needed to design safety into the vehicle, not just add safety on after a problem was discovered. This need drove NASA to become the home of some of the world's best design engineers and produced some of the best system safety, quality, and reliability engineers.

NASA demonstrated through the Mercury program that we could launch a human into orbit around the Earth and recover the astronaut and spacecraft safely. During the Gemini program, we perfected complex rendezvous and docking in space, and performed spacewalks. Both astronauts and equipment operated safely during longer durations in space. By the time the Gemini program ended, NASA was doing what was once thought impossible. Even with increasingly complex equipment and quick turn-arounds between missions, the astronauts always returned home safely. Success was becoming routine and expected.

NASA experienced a rude awakening in January 1967, when the Apollo 1 capsule burst into flames during a preflight ground test. The three astronauts performing the test perished in the blaze. The test had called for simulating a launch configuration, so the capsule was pressurized with 100 percent oxygen, and the hatch was sealed. Investigators determined that an electrical short sparked the fire. In a 100-percent oxygen environment, the fire quickly engulfed the capsule. But the test was being performed with an unfueled launch vehicle, so it was not considered hazardous! NASA never considered the possibility of a fire during the test—crew evacuation and fire suppression were not part of the test scenario.

NASA responded to this tragedy by strengthening safety oversight, clarifying responsibilities, improving communications, improving test safety analysis and emergency procedures, and making safety design enhancements to the Apollo spacecraft. Congress established the Aerospace Safety Advisory Panel to provide an independent review of the safety of NASA programs and operations. NASA established an Office of Flight Safety, independent of the flight program office, to review all aspects of design, manufacturing, test, and flight from a safety standpoint.

NASA recovered from this tragedy. NASA astronauts landed on the Moon six times and returned safely. The Apollo 13 mission demonstrated that NASA could recover from a serious technical mishap and return the crew safely to Earth. In the 1970s, NASA conducted the Apollo-Soyuz program and the Skylab program—logging more human spaceflight success.

For a period of time, America did not have a regular human presence in space. Throughout the 1970s, we were developing and building the next generation of [the] reusable space vehicle, the Space Shuttle. In the mid-1970s, Agencywide advocacy for flight safety became the responsibility of the NASA Chief Engineer.

From 1981 to 1986, NASA flew twenty-four Space Shuttle missions. Although we experienced some anomalies along the way, the astronauts always returned home safely.

Again, success was becoming routine—until a cold January day in 1986, when the Space Shuttle *Challenger* suffered a major failure in the seals of one of its boosters and exploded 73 seconds after liftoff. All seven crewmembers were killed.

In the painful months that followed, there were indepth, critical reviews by NASA and external bodies. The Shuttle program was grounded, and each safety practice was dissected and examined. Safety goals and procedures were revisited; even organizational and individual attitudes were considered. The reviews found a number of management flaws. For example, O-ring seal problems in the boosters had surfaced on previous missions. However, this information was not widely circulated. Concerns expressed by safety engineers did not always reach management in a timely manner. Additionally, the magnitude of the risk and the associated ramifications may not have been fully understood by the decision-makers. There had been growing pressure on NASA to launch the Shuttle regularly and on schedule. No one believed that they had enough data to prove that the launch was not safe. A collective mindset evolved—if no one could prove that the launch was unsafe, it must be safe!

In the few years after the *Challenger* accident, NASA put in place a number of improvements to its safety program. These included:

- Creating an independent safety organization, reporting directly to the Administrator.
- Increasing the budget and staffing for safety, reliability, maintainability, and quality assurance.
- Improving communications. NASA created an additional avenue to communicate safety concerns in a confidential manner—the NASA Safety Reporting System.
- Strengthening risk-management programs and initiating significant problem reporting, trend analysis, and independent systems assessment capability.

These improvements form the basis for today's Safety and Mission Assurance Program, and since return-to-flight in 1988, every NASA Space Shuttle flight has flown and landed safely.

How has human spaceflight safety advanced over the past forty years? Well, for one thing, we know more. We know more about engineering, materials, and robotics. Safety and mission assurance tools are much more advanced. We have the capability of improved nondestructive evaluation, and we can do computer modeling and sophisticated "what if" scenarios.

Today, we know more about program management and more about what it takes to fly safely. We know that there are a million things that can go wrong, and we know that we can never become complacent. We will not allow ourselves to be bullied by schedules, and we won't let cost constraints make us skimp on safety.

We don't ask our engineers and managers and experts to prove it is *not* safe to fly. Rather, we ask them to prove that it is safe.

This is a philosophical change from the days before *Challenger* and a fundamental management principle for safety of flight.

Today's human spaceflight safety prelaunch assessment review process is independent and comprehensive. For each launch, NASA managers prepare a Certificate Of Flight Readiness—we call it the COFR. Before I sign the COFR, I must personally understand all the safety issues and their resolution. If I do not have confidence that everything has been done to make the flight as safe as it can be, it is my job to *not* sign the COFR. The Administrator would not have it any other way.

The International Space Station heralds a new era of space exploration for America. On this program, safety is NASA's highest priority. My staff performs continuous oversight and independent assessment on the design, development, and operation of the International Space Station.

In sum, I'd like to describe the illustration shown [on page 74]. This picture represents NASA's safety hierarchy. We articulated the safety hierarchy a little over two years ago, as part of our quest to be the nation's leader in safety and occupational health, and in the safety of the products and services we provide. The safety hierarchy stresses that we are all accountable for assuring that our programs, projects, and operations do not impact safety or health for the public, astronauts and pilots, employees on the ground, and high-value equipment and property.

When people are thinking about doing things safely, they're also thinking about doing things right. And for the past couple of years, we've had some pretty good results. In the time since the failures of the Mars 98 missions that occurred in late

1999, every NASA spacecraft launch has met the success objectives, and every Space Shuttle mission has safely and successfully met all mission objectives. Now I can't say that NASA's safety program is solely responsible for these achievements, but, as we like to say, "mission success starts with safety."

In the future, looking forward, we will continue to make spaceflight even safer. That is NASA's vision. That is NASA's duty to both those who will travel into space and the American people who will make the journey possible.

What If? Paths Not Taken—John M. Logsdon

President John F. Kennedy speaks before a crowd of 35,000 people at Rice University on 12 September 1962. NASA Image 69-HC-1245.

I want to ask all of you to join me for a few minutes in a mental experiment. There is a certain sense of determinism as we review a period of history, like the forty years of U.S. human spaceflight. There is an implicit assumption that there were no alternatives to the way things happened. If you step back even half a step, you know that's not true; that along the way, history could have been very different if different choices had been made, if different events had happened. So I have arbitrarily picked a few situations in those forty years and invite you to ask along with me: "What if things had been different?"

This notion of counterfactual history has some legitimacy. I have used it as a class assignment for my students in space policy, asking them to write about what might have occurred if different choices had been made. Dwayne Day, a former student and now a colleague, has suggested a whole symposium on counterfactual space history, and that might be an interesting thing to do someday. As I looked into preparation for this talk, I discovered there is a body of literature on counterfactual history. And, not surprisingly in the Internet age, there are even Web sites devoted to the topic!

So let us start with the first "what if." The Mercury Redstone 2 flight on 31 January 1961 carried the chimpanzee Ham. It went too high and too fast. Ham experienced over 10-Gs on reentry, and the spacecraft landed several hundred miles down range. He was a very angry chimpanzee when rescue teams reached the Mercury capsule. The problem that caused the deviation in flight trajectory turned out to be very simple to identify; it was quickly diagnosed as a malfunctioning valve. It could have been fixed, and the next flight, which had been scheduled to carry the first astronaut, could have been launched without an intermediate test flight. But even in those days, safety was criteria number one. So Wernher von Braun and his team insisted on a flight of the repaired booster with a dummy spacecraft; that flight took place on 24 March 1961. The reality is if the 31 January flight had been successful, then the 24 March flight could have carried Alan Shepard. He would have been the first human in space, not Yuri Gagarin.

What might have been the impacts of that? It is reasonable to speculate that the Soviet reaction, the U.S. reaction to Yuri Gagarin's flight, President Kennedy's subsequent reaction to the Gagarin flight, the press reaction, and the political reaction that provided the fuel for Kennedy to ask his advisors to find a dramatic space program with which the United States could "win" might all have been entirely different. It is quite possible that the United States would not have decided to try to surpass the Soviet Union in spectacular space achievements. Then a very different space history would certainly have evolved.

Here is another possibility. In President Kennedy's inaugural State of the Union address, he invited the Soviet Union to cooperate in the exploration of space. In fact, early on, he had targeted space as an area for trying to develop mutual confidence and reduce tensions with our Cold War adversary. Kennedy was forced by the reaction to the Gagarin flight to compete, but he never gave up the cooperative idea. There's a book called *One Hell of a Gamble*[1] that traces the fact that Kennedy, between the

---

1. Aleksandr Fursenko and Timothy Naftali, *One Hell of a Gamble: Khrushchev, Castro, and Kennedy, 1958–1964* (New York: W. W. Norton and Company, 1997).

time he received the memo recommending Apollo and the time he announced Apollo on 25 May 1961, kept asking the Soviet Union "might you want to cooperate in space?" He received no response from the Soviets, so he went ahead with his speech on 25 May. Ten days later, in Vienna, he met Nikita Khrushchev for the one and only time and suggested "Why don't we go to the Moon together?" As Asif Siddiqi has suggested, at that point, the Soviet Union didn't have a lunar program, really didn't take the United States very seriously, and the official party line was to link cooperation to general and complete disarmament. So there was no positive response from Khrushchev.

Kennedy never really went away from the idea. In September 1963, at the United Nations in the most public possible way, he suggested, "Why should this be a matter of national rivalry? Why don't we do it together?" Khrushchev's son, Sergei, has written that at that point the Soviet leader was beginning to think more about cooperation. Kennedy, ten days before he was assassinated, sent a memo to Jim Webb asking for a plan to cooperate with the Soviet Union in space, including a cooperative lunar-landing effort.

What would have happened if Khrushchev's answer had been yes? Well, there are lots of possibilities. If the answer had been yes at Vienna in 1961, for example, the political support that made Apollo possible likely would have collapsed. This political support was based on competition, on the idea of the United States gaining a preeminent position in space. So if the Soviet Union had accepted Kennedy's offer, I'm not sure Apollo would have ever happened.

Could the Soviet Union have carried out its part of the program if cooperation had taken place? It is not clear whether the post-Khrushchev leadership of Brezhnev and Kosygin would have been as committed to this. It is also debatable whether the Soviet Union could have contributed to the program in the ways that would have made international cooperation possible.

Alternatively, if Kennedy had not been assassinated ten days after he had signed the government directive to find ways of cooperation, perhaps cooperation could have worked. Maybe the United States and the Soviet Union, the leading space powers in the 1960s, could have found a way to join forces. If that had happened, other things such as the International Space Station might have happened much sooner. It would have set a precedent for collaboration in space exploration which we are working on making succeed now. We might have been able to start down the cooperative path thirty to thirty-five years ago.

Here is another counterfactual notion to consider. Most of you are familiar with the 1969 recommendations of the Space Task Group that the U.S. accept a post-Apollo goal of manned planetary exploration before the end of the century and build a series of large space stations during the 1970s as steps toward that goal. What if, instead of rejecting that report out of hand in the aftermath of Apollo, Nixon said, "Yes, we'll do that." What might have happened? There is a fascinating book called *Voyage*,[2] by British engineer Stephen Baxter, that starts with exactly this

2. Stephen Baxter, *Voyage* (William Morrow and Company, 1997).

premise. The novel describes the first mission to Mars in the 1980s! It's a very enjoyable piece of counterfactual history.

If we had kept the Saturn V, if we had launched a 33-foot diameter instead of a 15-foot diameter Space Station, launched with one Saturn V flight by the late 1970s, where would we have been? What kind of Space Shuttle would we have built? If the Shuttle had been developed primarily as the supply vehicle for the station, we might have been able to build a fully reusable, straight-winged, highly operable vehicle. The Space Task Group report called for the initial mission to Mars in 1986 or, in the extremely ambitious von Braun version, 1982. We might have been at Mars by now if the choice to set that destination as a goal had been made over three decades ago.

We have all gotten used to the concept of an International Space Station. There was not a whole lot of debate leading up to President Reagan's late 1983 approval of the Space Station, of whether it should be international or not. The advocates of the international approach knew that there was opposition within the Reagan administration. So they didn't have it debated as part of the original decision package. The decision to make the station international came at high levels of the administration in the weeks before Reagan's announcement of the Space Station in January 1984. But what if the program had been a U.S.-only Space Station? What if we had not included international partnership? Would the station have survived its many budget and schedule problems over the past two decades? Howard McCurdy has written in his book, *The Space Station Decision*,[3]

3. Howard McCurdy, *The Space Station Decision: Incremental Politics and Technological Choice* (Baltimore: Johns Hopkins New Series in NASA History, 1991).

that approval of the station was a very marginal decision in terms of political support. When the station ran into problems in the 1980s and 1990s, without the international partnerships, I think the program would have been much more vulnerable and likely would have been canceled.

But there is an alternative possibility. The international character of the station program added cost and complexity to the program. If there had been no international involvement, the program might have gone forward, with NASA and its contractors settling on a feasible design that could be built within budget and on schedule.

What if the decision had been made to postpone *Challenger* because of the weather conditions on that January morning, and, when the mission was rescheduled and launched, would it have been successful? I think subsequent history would have been much different. Here are just some of the possibilities. Maybe there was an accident waiting to happen because of the attitude of increasing acceptance of risk. If the accident had not come on flight 51-L, it would have come sooner rather than later, and the consequences for the program wouldn't have been much different. Another possible scenario is that the Shuttle would have become increasingly reliable. The Shuttle would have continued to carry commercial and military payloads, not just NASA payloads. The plan at the time of *Challenger* was to launch twenty-four flights per year. We might have approximated that with adequate budget and improvements in reliability.

We would not have a commercial ELV industry if that had happened. Certainly the *Challenger* accident opened a window of opportunity. So maybe we would be using the Shuttle in a very different way than we are today.

Another possibility is that eventually the fixation on a Shuttle-only policy would have changed. We would have evolved into a more balanced and appropriate mixed-fleet strategy and be about where we are now.

Here is a final counterfactual possibility. The notion of inviting Russia to join the Space Station program has multiple parents, including Dan Goldin, Yuri Koptev and Yuri Semenov in Russia, and Leon Furth, who was Al Gore's National Security Advisor. There were many reasons to invite Russia into the program, but it was not a straightforward yes or no proposition. There was some significant skepticism about the wisdom of the idea. If Russia hadn't been included in the International Space Station program, what might have happened instead?

It may help to recall that in June 1993, the House of Representatives approved the NASA budget, including the Space Station, by one vote, 216 to 215. Bringing Russia in changed that to a hundred vote margin the next year. The Space Station was on the path to cancellation in the early years of the Clinton administration. It is thus a very plausible claim that bringing Russia into the partnership saved the station program, and, without Russia, it would have been canceled, and we would not have had to worry about all the problems with Russia as a partner.

Looking back at that period, the redesign team and then the advisory group to the White House, headed by MIT president Charles Vest, had several options that they looked at without Russian involvement. They believed that there were some good options for a station redesign. It is possible that we could have come out with a good station program, on budget, on schedule, if Russia had not been brought into the partnership.

A final possibility is that what has happened would indeed have occurred. As the United States and its existing partners found out how hard the station was and grappled with running the program as a multinational venture, including crew rescue vehicles and all the power modules that are required, the program would have ended up looking more or less as it has looked over the past decade.

I think the point of this exercise in counterfactual thinking is twofold—first, to recognize that not only have choices been made in the past that defined the character of what has happened and that different choices were possible and would have led to different outcomes, and, second, that we are currently making similar choices for the future. Today's choices obviously will have significant long-term consequences for space development. Decision-makers have an image of a desirable future when they make choices, but they also realize that the link between current choice and desired result is always uncertain. As the philosopher Yogi Berra is often quoted as having said, "making predictions is hard, especially when they are about the future."

Apollo and Beyond—Buzz Aldrin

Buzz Aldrin inside the Apollo 11 Lunar Module on 20 July 1969. This image was taken by Neil Armstrong, mission commander, before the two astronauts landed on the Moon. NASA Image 69-HC-893.

I got involved with spaceflight in a peculiar way. I graduated from West Point at a time when there was no Air Force Academy. I went in the Air Force at the time of the Korean War, and while there, I shot down a couple of MIGs. Years later, this led me to want to look at the extension of air travel into space. At MIT, I worked on intercepting other spacecraft.

Based upon that education, I got into the space program, not by route of the test pilot school. I was involved in a more esoteric, egg-headed approach. I did help to train the people who were on the first rendezvous missions.

I was slated initially on the backup crew for Gemini 10. That meant that I would skip two missions, and then I would fly on the prime crew with the next one. The only trouble was there was no Gemini 13. Because of a tragic aircraft accident that took the lives of the primary crew on Gemini 9, they had to make some crew adjustments. So Jim Lovell and I flew on Gemini 12. On that mission, I was able to take my SCUBA-diving expertise and training underwater for spacewalking and helped to teach some of the Navy people how to do spacewalks. Then, in the infinite wisdom of the Air Force, I was asked to command the test pilot school after I left NASA, even though I had never been through any test pilot training.

After awhile, I took what I did at MIT and developed the idea of cycling orbits. As an extension of rendezvous, I extended that to the situation of going to Mars. Finally I got around to writing a science-fiction story about going to the stars. I would recommend that you take a look at that because there's an awful lot of what I think we could be doing in the near future that is in that book.[1] In this story, the first human crews reach the moons of Mars in December 2018, fifty years after the crew of Apollo 8 reached the Moon. That is still a valid target that I think we can move toward.

Because of the cost of launching rockets, it looked to me as though there might be a need for some improvements in the Shuttle system. So I embarked on looking at fly-back boosters for the Shuttle. Those are pretty big machines, so we backed off looking at the Zenit as a reusable propulsion module inside of an airplane, and we started looking at Atlas 3.

I put together a company called Starcraft Boosters about four or five years ago. Then I also realized that a lot of people want to get into space. So I formed a nonprofit company called Share Space. I'd like to share some of my thoughts about what Share Space and our nation could do together.

Forty years ago, President Kennedy responded to Soviet space exploits by setting a course for the Moon. A similar bold stroke is required to answer the Soviet Union's inauguration of passenger travel to the International Space Station.

Passenger space travel is a huge potential market—big enough to justify the creation of reusable launch vehicles. Their low cost and high reliability will give the nation that develops them enormous commercial and military advantages over nations that continue to rely on today's space launchers.

1. Buzz Aldrin and John Barnes, *Encounter with Tiber* (New York: Warner Books, 1997).

Current systems fail 5 percent of the time on cargo flights and 1 percent of the time on crude flights, and they are very expensive, on the order of $5,000 to $10,000 per pound of cargo or people sent to orbit. Embracing passenger space travel, which leads to reusable vehicles, will reduce failure rates and costs by one or two orders of magnitude. The Share Space Foundation wants to ensure that America reaps the benefits of passenger space travel.

A near-term step is using the Space Shuttle to carry out economic and medical research on passenger space travel. Each year, more spare seats are available on Shuttle flights than on the Russian Soyuz missions. The Share Space Foundation would like to award these Space Shuttle seats to journalists and to winners of lotteries, auctions, and televised competitions.

As a nonprofit organization, Share Space would use part of the proceeds for research into the medical and training issues involved in passenger travel. This knowledge will speed the design of next-generation vehicles to serve the passenger market.

The rest of the proceeds might go into a fund to underwrite the cost of developing reusable launch vehicles. Reusable space transportation is the key that makes space affordable, enabling everything from expeditions to Mars to providing pollution-free electrical power to Earth. In addition, if the United States leads in the development of such vehicles, it will be the nation with effective control over the most militarily important high ground of the twenty-first century.

A little over a year ago, I worked with some people from the University of Texas, Massachusetts Institute of Technology, and the Jet Propulsion Laboratory to put together a paper for the American Astronomical Society entitled "Earth/Mars Transportation Opportunities, Promising Options for Interplanetary Space Transportation." We looked at cycling orbits and how we might build a spacecraft incorporating this concept. This was a three-cycler system, and we are in the process of doing an update on that paper for the American Institute of Aeronautics and Astronauts.

Now the space transportation elements that we're working on consist of two pillars. One of them is two-staged orbit, and it comes in three sizes—small, medium, and large. The other pillar of space transportation is Shuttle-derived cargo and habitable volume. That's based upon the external tank and tanks above that that would put up habitable volume.

For small boosters and orbiters, we could use the solid rockets on the Evolved Expendable Launch Vehicle (EELV) programs. We could also use a Russian engine, RD-120, which is the upper stage of the Zenit rocket. In orbiter small, there is an Air Force research lab, reusable upper stage, that they have in mind, and there's also an Air Force space maneuver vehicle that could be available and boosted with varieties of this booster small which, of course, can apply to the EELV program as it goes out this way.

The booster medium is what we would nominate as the star booster 200, and it has the unique property of having a rocket inside of it, the Atlas 3 rocket, with the RD-180 engine. Now the booster and orbiter medium could conceivably come on line relatively quickly and could provide crew-only transport for ten to twelve people to low-Earth orbit. Three starts could be made in this four-year period, this first administration.

The orbiter medium has a pod that can be ejected from the pad or from anywhere in flight. The essence of that ejectable pod and its capacity and its systems could also be used as a lifeboat, similar to the X-38. The orbiter medium, when boosted by one booster, goes into low-Earth orbit. With two boosters and a tank, it can then rendezvous with things at the L-1 port. The L-1 port really comes from the habitable volumes that are put up. We would envision looking at a prototype during this period and actually launching one before the end of the year 2008 into the space station orbit of the International Space Station, where it could supplement what we think is a desirable thing . . . an orbiter on station. Owen Garriott, who flew on Skylab, has been pioneering the activity of long-duration orbiters that could be left at the Station and relieved on Station by another orbiter, thereby relieving the burden of having to rely on the lifeboat Soyuz and a half module, both of which have been sort of postponed now by NASA because of cost overruns. The booster large now is a fly-back booster for the Shuttle, and two of those go with the Shuttle system as it proceeds toward phase out. One large booster launches an orbiter large into low-Earth orbit for Space Shuttle transportation two into the future.

With two boosters and a tank, it can then go to high orbits, which means it can intercept cycling space ships. Cycling space ships are a derivative of what we first put at the 51.6-degree inclination and then work close to the International Space Station, perhaps take the nose section of the tank and put it actually on the ISS as a larger half module than we plan to do right now.

The next series of habitable modules would go to 28.5 degrees and would eventually serve as Shuttle-derived quarters for astronauts. Shuttle-derived quarters for tourists eventually can become a lunar cycler, as we have two of these now in Earth orbit. We can be taking adventure travelers in the orbiter large now to these facilities. They've been going through a check-out phase here. The driving force for all of this architecture is high-volume traffic, which is required to make transportation of adventure travelers economical. You can't make a profit out of ten or twenty people once a month or once a week. You have to have sixty or eighty people every other day or more frequently.

Regarding the booster medium, there are existing elements such as the tank. There are also new elements such as an airplane that houses the Atlas 3 rocket. We have this sort of capability where we can replace the Delta class and the Titan 4 with upper stages boosted by the propulsion modules. It replaces the solid rockets on the Shuttle, and the fly-back boosters eventually replace the solid rockets on the heavy lift rocket that puts up both habitable volumes or two empty oxygen tanks or cargo that's going to the Moon or Mars.

We are also using Star Booster 200s with Atlas 3s in them, and we're using Castor 120s and a center stage. This will put 6 tons to geo-transfer, which accommodates most of the payloads going to geosynchronous orbits. If we want to match the Titan, then we use a larger booster here. It's launched vertically on the pad, and we obviously are not burning the solid rockets at liftoff. The boosters burn to fuel depletion at about Mach 5.5 or 6, so that the thermal protection system on the booster can be an aluminum heat sink. A few seconds after separation, the upper stage ignites and goes on and performs its mission.

The booster rocket is unpiloted, so naturally it will re-enter the atmosphere straight ahead. When it gets into enough atmosphere, it begins automatically turning back to the launch site. We showed this system to Bert Rutan, and it took him the longest time to realize that this was not taking off horizontally, but it was a vertically launched, unmanned vehicle. As I mentioned, this one booster can be mated with one orbiter of the same shape or of a shape of a derivative from the present Space Shuttle. We favor internal oxygen and external hydrogen for safety reasons and for the size of the orbiter vehicle . . . . The cruise back is about 200–300 miles, and we land robotically at 150 knots, and, when we're back on the runway, we then can remove the propulsion module and improve the reliability of the best rocket engines that are available.

When we have better rocket engines than the Russian RD-120s or RD-180s here, then we would institute these. We would expect that the fly-back booster might be built by Lockheed and might also have RD-180 engines on it, because I imagine that the orbiters from the space maneuver vehicle to the orbiter medium and the orbiter large that I envision would most likely be built by the world's biggest airplane company that acquired another airplane company, and then it acquired the company that built the present Shuttle.

It's been a pleasure for me to come back here on this day that honors one Shepard, and we're going to hear from another Shepherd a little bit later.

Breaking in the Space
Shuttle—Robert Crippen

Astronaut Robert Crippen, pilot of the first Space Shuttle mission (STS-1), in the cabin of the Columbia *orbiter during a mission simulation on 10 October 1980. NASA Image 80-HC-600.*

I like to look forward, but I am also a great student of history. I believe there are many lessons, both positive and negative, that we can learn by looking back. Quite often, we tend to forget some of those.

I'd like to speak a little bit about the era in which I entered the astronaut corps. I joined NASA in kind of a weird way—in the time period when everybody going into space was a test pilot. I was attending the Air Force Test Pilot School, even though I'm a Navy guy. In that time period, they were going through and selecting astronauts from the test pilot class, and I put my hand up and said, "I'd like to join." It turned out that both NASA and the Department of Defense were selecting astronauts, and, somewhere in the selection process, I ended up having to make a choice. There were lots of folks on the NASA list, and there weren't many folks on the Department of Defense list, so I figured that was my best chance to fly. So I said, "Send me to DoD for something called the Manned Orbital Laboratory," or MOL for short.

I did get selected for that program, which was canceled for various political reasons in 1969. During Apollo 11, when Neil and Buzz were on the Moon and Mike Collins was orbiting around the Moon, I and the other MOL crewmembers were interviewing with Deke Slayton saying, "Hey, you got any room for us in the program?" It turned out Deke said, "Okay. We could use some of you. But there's not going to be a chance to fly until around 1980 when we get the Space Shuttle done." At this time, the Space Shuttle wasn't even an official program. He said, "Well, I don't need all fourteen of you, though, so I'll take everybody that's thirty-five and younger." That cut us right in half, and I happened to be in the half that managed to go down to NASA.

When I arrived there, of course, NASA was right in the midst of the Apollo program. It turned out that we were already starting to cancel Apollo flights because of various financial and political reasons. But NASA had already moved a little bit out of the test pilot regime and picked two different groups of scientists. NASA administrators decided that scientists needed to learn how to fly. So these new scientists-astronaut trainees initially were sent off to the Air Force Training Command to learn how to fly and become pilots. The first of this group to fly was Harrison Jack Schmitt on the last Apollo flight, Apollo 17. Then there was a scientist-astronaut on each Skylab mission. Joe Kerwin, an M.D., flew on the first flight, Owen Garriott was on the second Skylab flight, and Ed Gibson was on the last.

So we were already starting to broaden the scope of what we wanted as far as crewmembers. There was a long hiatus between Skylab and the first Shuttle flight. The Shuttle was announced during Apollo 16, but it took a while to develop the Shuttle. We had one interim flight (Apollo-Soyuz mission), but we didn't have many people in the astronaut office. I think we got down to somewhere around twenty people. So it was pretty small, which I thought heightened my opportunity to get [to] fly—not like it is today with around 160 in the office.

But I was fortunate enough to be standing in the right place at the right time when the crew selection was made for the first Shuttle flight, and I was selected to be the pilot along with

Commander John Young. I always need to clarify that none of us red-hot test pilots wanted to be called a copilot. I was really the copilot, and John was the real pilot on the flight.

I was twenty-eight when I was selected to fly on MOL, but I was forty-three when I flew STS-1. So sometimes it does take perseverance. So if you have to wait a little while, it's not all that bad.

In 1978, three years before the Shuttle became operational, it was obvious that we would need a broader category of astronauts. The 1978 group was the first time that we selected what we referred to as mission specialists, who did not have to be pilots. We did not train them as pilots, but we did make them backseaters on the T-38, and, essentially, they all learned to fly.

That was also when we decided that we ought to have some women onboard. People keep talking about manned spaceflight, but it's better today to refer to it as human spaceflight. I was taught that very well on my second flight, STS-7, when Sally Ride was one of our mission specialists, and she kept correcting me if I said manned spaceflight. In addition to Sally, Kathy Sullivan happened to be on my last flight, and Kathy was the first U.S. woman to perform a spacewalk. So we broadened out who flies, the kind of categories we're looking for, and, as we move forward with the International Space Station, we do need more scientists rather than test pilots. This will be true even when we go back to the Moon and go on to Mars, which we certainly will, as Mr. Goldin said.

I will conclude by saying it is going to be interesting to see if somebody could pull together a symposium forty years from now and look where we're at in spaceflight. I hope some of the things that Mr. Goldin mentioned about the Moon and Mars will be realities at that time. It has been a privilege for me to have an opportunity to play a small part in that.

Going Commercial—Charles Walker

Charles D. Walker, STS-41D payload specialist, closes a stowage area for biological samples supporting the Continuous Flow Electrophoresis Systems (CFES) experiment in the Shuttle's middeck on 6 September 1984. NASA Image 84-HC-413.

Robert Crippen's comments take me back to my college days, when we were first introduced to spaceflight with the MOL opportunity. At that time, I was just beginning to think about what space might mean in my career, although I'd been interested in space for many years. I had read a lot of science fiction and closely watched Al Shepard's first short jaunt into space followed by Gus Grissom's.

Gus Grissom's flight actually brought it home pretty personally because Gus grew up 12 miles from my hometown. Twelve miles in Indiana is a long way, but still everybody is family within the state. So we thought of Gus as a hometown hero, and that impressed upon me that, gosh, real people can go into space. You have to be very well qualified and physically fit to withstand the rigors of spaceflight. But I figured that a small-town boy from Indiana could go into space if he gets a good education. So I went to Purdue University, the same school that Gus went to, and graduated in engineering.

Even then, I believed that I was going to design and build spaceflight equipment. It was only in the 1970s, as NASA began to build the first reusable space ship, that I came to the slow realization that, "It's not just the test pilots, not just the Bob Crippens of the world, that are going to get go into space, but scientists and engineers too. Gosh, I am an engineer. Maybe I can do this."

Lo and behold, I had the opportunity to go to work for one of the largest aerospace companies in this country, McDonnell Douglas. I specifically asked for the opportunity to work on a program that was going to fly equipment in space. The company at that time was subcontracting to Rockwell, which was the prime contractor to NASA for producing the Space Shuttle orbiter. The subcontract that McDonnell Douglas had was to build some of the propulsion equipment. So they said, "Okay. You want to build equipment that goes into space? Work on these tanks." That thrilled me for a little while, but then I said, "That's not really what I wanted."

I was looking for an opportunity to do engineering and scientific research with processes that could be commercialized. That's the point at which I wanted to jump into the perspective of the nonprofessional astronaut's opportunity to fly in space. Spaceflight is certainly a personal experience, and I'll touch on that.

But I also want to reflect first on the professional experience. My opportunity to fly came through my employer at the time. McDonnell Douglas was looking for an opportunity to exploit the unique microgravity environment of space. In the late 1970s, the company selected a process called electrophoresis, which demonstrated the capability to purify biological materials. Electrophoresis was first very briefly demonstrated on the Apollo 14 mission. In fluid form, virtually every biological substance is part of a very complex mixture of proteins and cellular materials. Electrophoresis is useful to purify pharmaceutical and other chemical products. Purification processes here on Earth have to operate within gravity. But for separation processes in biotechnology, we see impurities created because of the churning in an imperfect purification process. Going into a low-gravity environment should prevent that from happening, and that was the theoretical basis for the electrophoresis technology.

The early experiments didn't seem to validate that, but we wanted to carry it forward on a scale that had not been seen before. The Space Shuttle was the obvious way to do that. It was the space truck. It was going to be going into space with plenty of cargo capability, with crew attendants to monitor and modify the early experiments or, later on, the production processes. So McDonnell Douglas applied to NASA for the opportunity to fly the experiment. The first opportunity was on the Spacelab module. Unfortunately, it looked like NASA's process of reviewing our equipment and our processes for safety and function was going to take up to three years.

In the business world, that's called a showstopper. When you've got one-year research budgets, you have reviews even more frequently, and your budgets are scrubbed one year at a time. You have to show results. We needed to have some real significant progress and creative thinking in a lot less than three years' time. This was before John Young and Bob Crippen's first mission, the first flight of STS.

The decision was made to try to fit the experiment package into the crew compartment. Later I found out that this caused some friction because our package was going to infringe upon the food preparation facility space. We accepted that opportunity, and we engineered the experiment to fly in that pressurized crew compartment volume, rather than in the research module, and to be tended by the crew.

The first time it happened was on STS-4, in June 1982, and this flight was very successful. But I want to point out that it took a big investment on the corporation's part, since this was the first such NASA-industry joint endeavor agreement in which there was no exchange of funds. NASA got from the arrangement the opportunity to do research with our equipment. A third of the time on orbit was for NASA-sponsored research, and the remaining two-thirds was our time. For that opportunity and for the integration of our equipment into the Space Shuttle and for crew training, McDonnell Douglas invested millions of dollars.

The safety and fit checks took up to a year-and-a-half to go through. Now this was the first time this had been done aboard the Space Shuttle inside the crew compartment. There had been outfitting for experiments in the Spacelab module, but this was different. The process was grueling at times. A lot of engineering and technical effort was put into meeting the safety and fit requirements.

Function was different. It didn't necessarily have to function. We didn't have to be successful, except for our own objectives. NASA simply wanted to make sure that it was safe for the Shuttle vehicle, safe for the crew, and that it fit in place properly.

We did fly that first time in June 1982. What was unique about that, from a historical standpoint, is that not only were we the first industrial payload to fly aboard the Space Shuttle, but we also did it in an extraordinarily rapid period of time. That year or so to prepare for that first flight was then followed by six more flights within forty-one months. There is perhaps only one other industry-sponsored or industry-participating space payload that has had such a rapid turnaround time. That's flying once every seven months with a different research objective. Most of the equipment was the same, which is what allowed us to fly so rapidly. Industry remained interested because of the very rapid turnaround time and frequent access that allowed an evolution toward a research and a commercial objective. You've got to have that research environment where you can test. You'll scratch your head over the results, get to that "ah-ha" moment, and then go back and try it again as quickly as possible.

Reflecting back to that flight in June 1982, the experiment operator was one of the two crewmembers onboard on that mission. The first four missions only had two crewmembers onboard. Hank Hartsfield, in the right seat on that mission, was the principal operator of the equipment. He turned it on and off, and took some very good photographs that helped us evaluate the experimental work.

It went very well, so much so that then my company asked NASA the big question. "Say, training mission specialists and pilots to conduct our research is good and important, but the best way for us to obtain the maximum result is to send a research engineer along." NASA's response was, "Well, we've kind of been looking forward to an opportunity like that. We don't know that now is the time, but now that you've asked the question, let's see if we can find an answer. Maybe yes, but they've got to meet all the qualifications that our career astronauts do." NASA then asked my company, "Have you got anybody in mind?" Well, lo and behold, I'd had my hand up for two or three years already. It was getting tired by that point in time, so I was glad that NASA said maybe yes.

It took a while to go through the checkout and qualification process, and, quite frankly, I didn't go through the routine selection process at the same time that the astronaut candidates were going through selection; although the process was the same. So it was a bit of a challenge to synch up with the Astronaut Office since

I wasn't even a resident there. I was still a resident in St. Louis at the time. NASA worked with me and the company on schedules and travel to and from Houston, and it worked out very well.

The opportunity was absolutely marvelous. It was a personal and a professional challenge absolutely. But the opportunity to fly with not only these tremendous career NASA astronauts, but to go into the environment that few other human beings have had a chance to do was a thrill beyond belief. To accomplish some career objectives in that regard made it even more exciting.

Since then, times certainly have changed. When talking about research in space, and in particular the industrial perspective, we've got to come forward from that time in the early 1980s to today. Today we think about the Space Shuttle in the form in which I daresay it was originally intended, as a space truck, not as a laboratory going into space and coming back, but as a truly versatile space transportation system.

We're building an International Space Station in orbit. Sixteen international partners are putting together a research laboratory that is unsurpassed in that environment. Just getting it there is the largest technical peacetime program ever undertaken. The challenges are immense, but the opportunities are tremendous too. Science is being done there now and will continue to be done. Industrial research will be done there as well, both in basic scientific and technological applications.

There have been some recent studies such as the KPMG marketing study for commercial opportunity and commercial applications on the ISS, a study which was done for NASA in 1999. Their finding was that the commercial opportunities were still very limited as seen by commercial industry. Industry still sees two high hurdles: cost and frequent access. Again, industry needs to respond to boards of directors and to stockholders, who need to see results. You need to have action. You need to respond to research and objectives, and to test these objectives in a potentially commercial program and know where you're going. You need to be able to change course quickly if necessary and cut losses when necessary.

Right now to get on the ISS, it's about an eighteen-month cycle for review of safety, fit, and functionality. That's just for the first flight aboard a research rack. That's a little daunting in itself.

But unfortunately we are still largely at the point where you say, "Well, when I can redo my experiments, how quickly can I readdress them?" And the response is, "Well, you've got to be able to either have telepresence installed such that you can manipulate conditions, temperature, pressure or depending on what your experimental objectives are and your equipment's capability, or maybe we can find some crew time, and we'll be able to arrange that, but probably not on the schedule that you would like instantaneously, or you'll have to wait until the equipment or some part of it can be brought back home or replaced in some many months, or maybe even a little more than a year's time." It is very difficult for industry to accept limitations of those kinds.

Another issue is that there are no payload specialists coming up through the ranks anymore. That was my category of noncareer astronaut, a payload specialist, one who specifically focused on a research mission. There aren't any in the program today, and I think that's a problem for industry and for science and research.

In closing, while the conditions are more rigorous today for the ISS than they were in the very early days of space travel, opportunities still abound, and we just need to overcome the hurdles. As Pogo put it, "By gosh, we seem to be surrounded by an insurmountable opportunity here." This really is a great time in human spaceflight. We're doing marvelous things up there from an engineering standpoint. We now have to put them to good use. We need to optimize the 30 percent of the ISS that our federal government and the international partners have available in terms of the Station's power, volume, and crew time. Despite the recent issues with cost and schedule, as Mr. Goldin has said, this Agency will find a way.

This country and the partners will find a way to restore the ISS's capability. We need help from this government, from our Congress, from our partners to do that, but it will be done, and then this facility is going to be world class—nah, it will out-of-this-world class.

I'm pleased to be a part of not only the history of spaceflight and the history of industry's participation in spaceflight, but I'm also pleased to be a part of the future, the future applications, the future benefits that our spaceflight program is going to bring to our economy, to our careers, and to those of us that are both taxpayers and participants as well, to the great joy of seeing success as part of this country, as a part of our intellect, applied to the great beyond.

It's a wonderful tomorrow. I'm glad to be part of it.

Science in Orbit—Mary Ellen Weber

Mary Ellen Weber works with a syringe related to the Bioreactor Development System (BDS) aboard STS-70 in July 1995. NASA Image 95-HC-486.

Imagine, as a scientist, having the opportunity to study an environment that literally exists nowhere on Earth, and then imagine actually being one of the test subjects. The human body has evolved over millions of years under the force of gravity. Every system in your body—how your fingernails grow, how you digest food, how you circulate blood—has evolved to use gravity.

The reason your heart is not in the middle of your body, but in the upper part of your body, is because of gravity. We take ourselves as test subjects into space in an environment where there is no gravity, and we get to observe what happens. All the sensors that tell your body how much water you should have are in your neck, and, when all that fluid shifts up to your neck, your body thinks you have too much water. In space, astronauts function with about 30 percent less water than they do here on the ground. That's just one of the amazing changes that takes place in the human body. As an astronaut, you get to see this happen to your body, to see your body struggle with it, and then, in just a matter of days, you get to see your body adapt and live and flourish in this whole new environment.

This was an opportunity that I could not even imagine initially. I could dream about it when I was in college getting a chemical engineering degree and then later getting a doctorate degree in chemistry.

I had a dream to go in space and to be able to experience this and be able to make these kinds of studies and observations. But I only hoped; I didn't think it would actually happen. When it did, back in 1992, it was certainly a dream come true. When I finally had the opportunity to fly in space in 1995 for the first time, it was just the most incredible thing I have ever experienced.

Now most people ask me, they say, "Well, what is your specialty as an astronaut?" Well, the science astronauts, the mission specialist astronauts, don't actually have a specialty. We do our science vicariously through the other principal investigators that work year after year, dedicating themselves to designing experiments, to asking the critical questions that the astronauts can then answer when they're up in space.

One of these experiments that I got to fly with back in 1995 was the bioreactor. This is an incredible experiment and incredible equipment. What it allows you to do is grow not just cells, not just a layer of cells in a petri dish, but to grow human tissue outside the human body. I had the opportunity to fly with the first bioreactor of this new generation in 1995 and watch colon cancer tissue—not cells, but tissue—grow before my eyes.

Charlie Walker discussed the challenges of commercializing technology. Since his firsthand experience, I have been able to work with a venture capital firm which has decided to invest in this technology. There is a market need out there that can compensate for the high cost and the limited access of space, and we believe we have a winner here.

It is just an amazing opportunity, as an astronaut, to be a part of both flying the experiments in space and of helping commercial companies identify the opportunities that space brings. This company is trying to make a liver-assist device. Right now if you have liver failure, you have two choices: you either get a transplant or you die. There is nothing else out there right now. But with the bioreactor that can grow tissues that function like the tissues in our body, we believe we can make a device that can save millions of lives. At the heart of it, one of the key elements of that is space.

Another feature of space is that it is a very quiescent environment, which allows us to obtain structures of proteins and design new drugs. This is not a pipedream. This is something that is happening and has been going on for a decade. There's a new flu medication coming out that is targeted to a very specific flu enzyme that keeps any flu virus from attacking your body, and, because we know the protein structure, we're able to design this drug with very limited side effects.

This is what is possible when you have a whole new environment. These are the possibilities of science and space. We have just begun to tap into this with weeklong flights or two-week flights, and we are only beginning to chip away at some of these questions. With the Space Station now, 365 days out of the year we have an opportunity to enter into a whole new era of space science—not just understanding space, but seeing how human bodies, proteins, and tissues react and grow in space.

I am so proud to be a part of this space program, so proud to be a part of something bigger than all of us. For thousands of years, people looked up at the sky and tried to imagine what was out there, what were those points of light? They made up stories about astrological figures. They just couldn't imagine it all. We are so fortunate in these past forty years to be alive at the time when we are just beginning to get the answers. It's not only amazing to be an astronaut, it's an amazing time to be alive. With the sacrifices and the commitments of so many people in this country, we will indeed propel our civilization into this next era of space exploration.

Training for the Future—T. J. Creamer

and, the truth is, we've got a great mission and a great bunch of people working together. It's an exciting time to expand that horizon that the literary minds helped expand all along; now we get to go a little bit further with our feet.

In the not-too-distant future, we will have completed the Space Station. Then the question becomes where to go next. Of course, as you've heard this morning, some of the likely targets are the Moon and Mars.

It is the giants I mentioned earlier upon whose shoulders we stand. As it turns out, Bob Crippen's middle daughter helps carry this legacy forward. I walked into class, and his daughter was our primary instructor. She made cookies to keep us interested in the class and teased us with candy, and threw it at us when we got the right answer.

Then she took us into the simulator and showed us how many ways we can do bad things to ourselves. She is a very sweet, smiling, and wonderful mentor, but the truth is that she is helping bring us into the future so we can do our job and become part of the legacy that our predecessors formed for us.

The truth is we are extremely excited about what we are doing. I can't wait to fly, and the flying is only a small portion of the job. We are making contributions every day in building the Space Station and helping to build the future.

*Expanding the Frontiers of
Knowledge—Neil de Grasse Tyson*

A spaceborne radar image of the Great Wall of China in a desert region about 400 miles west of Beijing. The wall is easily detected from space by radar because its steep, smooth sides provide a prominent surface for reflection of the radar beam. The image was acquired by the Spaceborne Imaging Radar-C/X-Band Synthetic Aperture Radar (SIR-C/X-SAR) aboard the Shuttle Endeavour *on 10 April 1994. NASA Image 96-HC-228.*

I do not want to preach to the choir today. You are the choir, and we have a panel of preachers here. Instead, I want to try to stir the pot a little and maybe get you angry, but hopefully not with me. I want to offer a point of view that I have not seen discussed in forums such as this. It is a point of view that represents the landscape in which most of this conversation would need to be present if any of our grand plans for space exploration are to come true.

We all know how important it is to fly. We have been doing it since 1903, and it is a point of pride in our species. We finally could do things the birds could do.

By the way, many people before the Wright brothers said that it would be impossible to fly. Even some well-known scientists, Lord Calvin among them, said, "We'll never have heavier than air flight." Well, of course, birds are heavier than air, and they have no problems flying. So there couldn't be a law of physics against heavier than air flight. So we have to be careful about what we say is impossible.

I would like to offer an argument for why I think it is unlikely that humans will ever leave low-Earth orbit for the next hundred years. I am not going to invoke laws of physics to make that point, because, basically, if the laws of physics do not forbid something, then it can happen. This is why, of course, the people who said nothing would ever go faster than the sound barrier were ignorant, because, in fact, we already had rifle bullets traveling faster than sound.

So when I say, no, I do not believe we will send people out of low-Earth orbit, it is not because I do not think it is scientifically possible. That is no longer an issue. I believe we have enough confidence in our technologies today so that basically anything we say we want to do, we believe we can do it. There is no doubt about it.

I have done a study of the major funded projects in the history of the world. I think we could quibble about what major means, but I can provide a short list of five or six that we would all agree are some of the major funded projects ever conducted. One of them would be the pyramids. Another one might be the Great Wall of China. The Manhattan project and the Apollo project were two other major funded projects where significant resources of a nation were redirected to enable these projects to take place. I would also include great explorations such as the voyages of Columbus and Magellan. We might imagine that going to Mars or going anywhere other than low-Earth orbit might constitute a major funded project in modern times.

In looking at the history of such projects, I found exactly three drivers. Only these three incentives can induce a population to agree to spend that much money on a major project.

One driver is, of course, economics. If there is a promise of economic return, it is an investment. Such were the voyages of Columbus and of Magellan, and of most explorers of that era.

Incidentally, if you went to the explorer himself and said, "Why are you doing this?" He would typically say, "Oh, because I'm an innate explorer. I want to know what's on the other side of that mountain." These are admirable traits, and the leaders of such expeditions have those traits. But in the end, somebody behind the explorer is writing a check. Italy would not write a check for Columbus, but Spain did.

A second driver is the gratification of ego. That ego can include the praise of a leader or a deity. So major funded projects such as the pyramids were basically elaborate tombstones. The Taj Mahal and the Vatican were built to praise something powerful.

There is a third cause. It is perhaps the most obvious driver—military defense. As examples, the Great Wall of China and the Manhattan project were clearly motivated by national security concerns.

A fascinating combination of economics and defense is the interstate highway system. After Eisenhower came back from the European theater of operations where he saw how effectively Germany was able to move goods and services over the Autobahn, he said, "I want some of those back home." So the interstate system was built for economic reasons, as well as national security.

So these are the three drivers that I have found. I have not found one exception to these.

Now let's go back to the Apollo program for a moment. I remember the 1960s. I knew from age nine that I wanted to study the universe. I watched the Apollo astronauts land on the Moon, and it meant something special to me. But when I went home, I encountered other things, such as picketing outside of the apartment building where my family wanted to move into to prevent it from integrating racially. There are other real-world problems such as the next paycheck that affect the general population.

Also, I remember that there was this mood surrounding the space program, "We are Americans. We're explorers. We're going to explore space. Let's go ahead. This is natural for us." The film we saw this morning was honest about this competition with the Soviet Union. But let's be a little more blunt about it. The most famous, resonant lines ever to come out of President Kennedy's mouth reflect this international competition.

Spoken 25 May 1961: "I believe this Nation should commit itself to achieving the goal before the decade is out of landing a man on the Moon and returning him safely to the Earth. No single space project in this period will be more impressive to mankind or more important for the long-range exploration of space, and none will be so difficult or expensive to accomplish."

If you hear that, you say, "By gosh, America is about exploration. We're about the exploration of space. That's the next frontier—just like the Columbus voyages and all the great explorers of the fifteenth century. Our next new ocean is space." But remember that this speech was given only six weeks after Yuri Gagarin was launched into orbit by the Soviet Union, a second jolt to the American ego after Sputnik in 1957.

The following paragraph of the speech reads, "If we are to win the battle that is now going on around the world between freedom and tyranny, the dramatic achievements in space which occurred in recent weeks should have made it clear to us all, as did Sputnik in 1957, the impact of this adventure on the minds of men everywhere who are attempting to make a determination of which road they should take."[1]

---

1. Special Message to the Congress on Urgent National Needs, 25 May 1961. *Public Papers of the Presidents of the United States: John F. Kennedy, 1961* (Washington, DC: U.S. Government Printing Office, 1962), pp. 396–406.

Now if that's not a military speech, I don't know what one is. So that's where the check writing happens, because we view this as a threat. It's a military decision. Kennedy was a big dreamer, but, in the end, one of these three causes reared its head. It happened to be defense couched on the premise of exploration.

Now, of course, space travel is expensive. We like to believe that one day it will be cheap. I'm not really all that convinced of it. I have the Space Shuttle operations manual read by Shuttle pilots in 1982. The whole opening section talks about how cheap it will be to run the Space Shuttle, how much cheaper it will be than any previous way we ever went into orbit. Entire Space Shuttle missions would cost $30 million, tops. That's not what it turned out to be, of course. It costs $200 million a day to keep the Space Shuttle in orbit. Now you can say, "Well, it's cost overruns," but my concern here is not even so much how much it costs, but whether a nation can sustain that through the ebbs and flows of political climate and the ebbs and flows of economic cycles.

If we can send people to Mars for $50 billion, let's do it. Evidence shows that if you start out saying it's going to be cheap, it just does not end up that way. So we need a realistic plan here.

When one of these three drivers is in effect, cost doesn't matter. Yes, there were debates in Congress about the cost of the Apollo program, but they were just pro forma. We spent whatever it took to put Buzz Aldrin and Neil Armstrong on the Moon.

Now is space really our frontier the way the oceans were the frontiers of the old explorers? Space, as you know, is supremely hostile to life. You can say, "Well, explorers had it bad too." In the 1540s, Pizarro led an expedition into South America to look for the fabled land of oriental spices.

He sailed across the Atlantic to come to South America. He had with him 4,000 men, hundreds of cattle, horses, dogs, a virtual moving city-state, in search of this land. He had it hard moving across the mountain ranges and in the valleys and in the rain forest.

I will quote from William Prescott's account of this: "At every step of their [the crew's] way, they were obliged to hew open a passage with their axes, while their garments, rotting from the effects of the drenching rains to which they had been exposed, caught in every bush and bramble, and hung about them in shreds. Their provisions, spoiled by the weather, had long since failed, and the live stock which they had taken with them had either been consumed or made their escape in the woods and mountain passes. They had set out with nearly a thousand dogs, many of them the ferocious breed used in hunting down unfortunate natives. These they now gladly killed, but their miserable carcasses furnished a lean banquet for the famished travelers."[2]

On the brink of abandoning all hope, Pizarro said "I'm going to send half of you guys back. We're not finding the place. It's probably still out there, but we're running low on supplies. I'm going to send half of you back to get more supplies."

So he sent one of his top men back to get more supplies.

2. William H. Prescott, *History of the Conquest of Mexico and History of the Conquest of Peru* (New York, Random House, Inc., undated), p. 1,074.

But how did he do that? How did he sail the river? The forest furnished them with timber. The shoes of the horses, which had died on the road and then had been slaughtered for food, were converted into nails. Gum distilled from the trees took the place of pitch, and the tattered garments of the soldiers supplied a substitute for oakum.

At the end of two months, a brigantine was completed, rudely put together, but it was strong enough to carry half the company. The company got on the boat, went down the river. The river fed into the Amazon. They said, "Hey, the Amazon, we know where this goes. It goes into the Atlantic." They went to the Atlantic and then sailed back to Europe. Pizarro, after waiting for these guys, found out they weren't coming back. So then he hiked his way back to their base camp and went back to Spain.

My point here is they came to South America and bad things happened, so they just went home. Yes, it was costly to life and limb. But I wonder what would happen if we would send people out to space and suppose they crash-landed on a planet.

If we accept that space and the oceans are analogous frontiers, what would such marooned astronauts do? They would have to mine for new materials on the surface of the planet and rebuild their spacecraft. They would need to extract silicon from rocks and make silicon wafers and imbed circuitry; rebuild their computers. Then they would need to relaunch themselves back into space. That is the counterpart to the ocean-going exploration analogy.

Of course, if astronauts crash-landed somewhere unknown, the planet might not have air. At least Pizarro's team had air to breathe. They did not have to worry about whether there was oxygen.

So space is supremely hostile, but we know this. But when we ask what is the cost of human space missions, we need to consider as many contingencies as possible. This is important because we want to do more than send people on one-way trips, we want to be able to bring astronauts back.

So if exploration is what really matters and not just pride of nation, then perhaps we should genetically engineer a version of ourselves that can survive the hostile environments of space. We've got cloning. We're inside the genome. Let's just do it. Well in fact, we've done that already. Yes, we have emissaries of ourselves that survive the hazards of space; they're called robots. You don't have to feed them or bring them back, and they don't complain if you lose them in space.

So my concern is if costs turn out to be what they have historically been and the time to execute programs lasts as long as it historically has, then I am not convinced that economic cycles and political cycles will allow such programs to survive if they do not satisfy one of these three criteria. The record of history tells us this, unless somehow you want to believe that we are different today than 6,000 years of our predecessors.

Pushing Human Frontiers—Robert Zubrin

An artist's conception of two habitats that a crew would connect while exploring Mars. This image was produced for NASA by John Frassanito and Associates. NASA Image S93-45581.

Dr. Tyson offered a very interesting set of ideas. I agree with some of the points he made and disagree with others. But I'm here to talk not about how we're going to mobilize the political or technological forces to get us into space, but rather why we need to do it.

I'm going to start out with a quote by a very eminent historian. Frederick Jackson Turner gave a presentation entitled "The Significance of the Frontier in American History" at the annual meeting of the American Historical Association. Incidentally, this was three years after the frontier was declared closed in the American census of 1890. He was looking backwards on 400 years of European presence in the Americas. Turner wrote:

> To the frontier, the American intellect owes its striking characteristics. That coarseness of strength combined with the acuteness and inquisitiveness; that practical, inventive turn of mind, quick to find expedients; that masterful grasp of material things, lacking in the artistic but powerful to effect great ends; that nervous buoyancy, the energy, the dominant individualism, working for good and evil, and withal that buoyancy and exuberance which comes with freedom—these are the traits of the frontier, or traits called out elsewhere because of the existence of the frontier. Since the days when the fleets of Columbus sailed into the waters of the New World, America has been another name for opportunity, and the people of the United States have taken their tone from the

incessant expansion, which has not only been open but has even been forced upon them. He would be a rash prophet who should assert that the expansive character of American life has now entirely ceased. Movement has been its dominant fact, and, unless this training has no effect upon a people, the American energy will continually demand a wider field for its exercise. But never again will such gifts of free land offer themselves. For a moment, at the frontier, the bonds of custom are broken and unrestraint is triumphant. There is not tabula rasa. The stubborn American environment is there with its imperial summons to accept its conditions; the inherited ways of doing things are also there; and yet, in spite of environment, and in spite of custom, each frontier did indeed furnish a new opportunity, a gate of escape from the bondage of the past; and freshness, and confidence, and scorn of older society, impatience of its restraints and its ideas, and indifference to its lessons have accompanied the frontier. What the Mediterranean Sea was to the Greeks, breaking the bond of custom, offering new experiences, calling out new institutions and activities, that, and more, the ever-retreating frontier has been to the United States directly, and to the nations of Europe more remotely. And now, four centuries from the discovery of America, at the end of a hundred years of life under the Constitution, the frontier has gone, and with its going has closed the first period of American history.[1]

So Turner's basic point was that the fundamental American character, our philosophical outlook, and our forums and institutions were all based upon the existence of the open frontier. He documented this at great length in his book and also the fact that many of the key issues in the growth of America and the key turning points all hinged on the frontier.

The question that he raised explicitly regarding the end of the frontier was what happens to America and all it stood for? Can a free, innovating society be preserved in the absence of room to grow? Turner predicted a growing bureaucratization of American society, increased hostility to immigrants, increased skepticism on the idea of progress, and a decrease of the ability of both institutions and individuals to take on risk, and other associated social phenomena. With respect to the issue of bureaucratization, he seems to have been on the mark.

If we want to have the kind of freedom that Americans had prior to the closing of the frontier, a new frontier is required. Now one could discuss where such a new frontier might be—Antarctica, the oceans, the Moon, asteroids, or orbiting space colonies. At this stage of human history, I do not believe that any terrestrial environment can afford the function that a true open frontier did in the

---

1. Frederick Jackson Turner, *The Frontier in American History* (New York: Holt, Rinehart, and Winston, 1947, 1962), pp. 37–38.

past. Simply put, wherever you are on Earth right now, you are within convenient range of communication and transportation technology. There is no new place on Earth where a new branch of human civilization can actually develop. Another way to understand this is if the American Revolution had happened today as opposed to 1776, the colonists would have lost. The colonists only were able to break away from Britain and go their own way because of the extreme logistical difficulties associated with maintaining control across transoceanic distances in the eighteenth century, because the British clearly outnumbered us.

So if you want to create a truly new and independent branch of human civilization that can experiment in new forms of existence and go its own way, it does have to be in space. Without going into detail, I believe that of all the places within reach of our technology, Mars is by far the best prospect because it is the planet that has all the resources needed to support life and, therefore, potentially civilization. By contrast, the Moon does not have these resources. Mars has got what it takes. It is far enough away to free its colonists from intellectual legal and cultural domination of the old world and rich enough in resources to give birth to a new civilization.

Now why do we need to go to Mars? Why do we need, more generally speaking, a new frontier in space? I believe the fundamental historical reason is because Western humanist culture will be wiped out if the frontier remains closed. Now what do I mean by "humanist culture?" I mean a society that has a fundamental set of ethics in which human life and human rights are held precious beyond price. That set of philosophical notions existed in what was to become Western civilization since the time of the Greeks, the immortality and divine nature of the soul as popularized by Christianity, but it never became effective as the basis for ordering society until the blossoming of Christendom into Western civilization as a result of the age of discovery.

An artist's conception of a vehicle to help humans explore Mars. This image was produced for NASA by John Frassanito and Associates. NASA Image S93-050645.

The problem with Christianity, despite its very interesting philosophical notions, was that it was fixed. All the resources were owned. Basically, it was like a play where the script had been written, and the parts had been assigned. There were the lead players, the bit players, the chorus, and there was no place for someone without a place. The new world changed that by supplying a place in which there were no established ruling institutions—a theater with no parts assigned. The new world allowed for the development of diversity by allowing escape from those institutions that were enforcing uniformity.

There are many problems that face us before humans can actually land on Mars, but there is fundamentally no comparison with our situation forty years ago. We are much better prepared today to launch humans to Mars than we were to launch humans to the Moon in 1961. We are better prepared technologically, scientifically, and financially. We have no credible military opponent who commands our resources. It is true that the fact that we had a military opponent did put a little drumbeat on things, but, from a material point of view, we are certainly better prepared in every respect.

Let us talk about what the twentieth century might look like without a Martian frontier. For one thing, I think we would be looking at declining human diversity. Global communications and jet aircraft are linking the world together very intimately, and so cultural diversity will, of necessity, decline. In biology, an animal type is considered strong if it has many diverse components, and I think it's ultimately a weakening of human society that we are faced with losing diversity. However, the same generic level of technology, which is making impossible the maintenance of diversity on one planet, has now opened up the prospect whereby new and more profound levels of diversity can establish the expansion of new branches of human society on other planets.

I also believe that without a new frontier in space, we face the risk of technological stagnation. The B-52 airplane is emblematic of technological stagnation. The B-52 went into service half a century ago, and it is still in service. It would have been inconceivable to any of the pilots flying the first B-52s that their grandchildren serving in the United States Air Force would be flying the same aircraft. Technological progress has actually slowed down in the last portion of the twentieth century. There has been obvious progress in certain fields such as computers and electronics. But if you look at society overall, in the first third of the century, we went from the horse and buggy world to a world of automobiles, telephones, electrification, and radio. Aviation went from the Wright flyer to the DC-3. In the second third, from 1933 to 1966 or so, we went to color TV, nuclear power, jet fighter aircraft, and Saturn V rockets. If we had continued on that vector, today we would have ocean and Moon colonies, solar-powered cars, and fusion reactors, but we obviously do not. The world today, at least in terms of advanced technology, has not changed that much since the late 1960s, especially compared to how it changed in the previous thirds of the century.

A frontier is a tremendous driver for technological process, because what you typically have at a frontier is a labor shortage. One of the most wonderful things about colonial and nineteenth-century America was the tremendous labor shortage. Despite everything you've heard, which is all true, about the horrendous conditions in the industrial revolution in New England and such places, the fact of the matter is that wages there were vastly higher than they were in similar establishments in Europe, and that's why millions of people voted with their feet to come here. At every level of society, opportunity was better here. Furthermore, because labor was so expensive, there was this tremendous driver for technological progress, for the creation of labor-saving machinery.

On twenty-first-century Mars, no commodity is going to be in shorter supply than human labor. There is going to be a tremendous drive for advanced technology, some of which might be otherwise blocked on Earth because of popular concerns about the environment or genetically engineered crops.

With human colonization of Mars, I think you will see a higher standard of civilization, just as America set a higher standard of civilization which then promulgated back into Europe. I think that if you want to maximize human potential, you need a higher standard of civilization, and that becomes an example that benefits everyone.

Without an open frontier, closed world ideologies, such as the Malthus Theory, tend to come to the forefront. It is that there are limited resources; therefore, we are all in deadly competition with each other for the limited pot. The result is tyrannical and potentially genocidal regimes, and we've already seen this in the twentieth century. There's no truth in the Malthus Theory, because human beings are the creators of their resources. With every mouth comes a pair of hands and a brain. But if it seems to be true, you have a vector in this direction, and it is extremely unfortunate. It is only in a universe of infinite resources that all humans can be brothers and sisters.

The fundamental question which affects humanity's sense of itself is whether the world is changeable or fixed. Are we the makers of our world or just its inhabitants? Some people have a view that they're living at the end of history within a world that's already defined, and there is no fundamental purpose to human life because there is nothing humans can do that matters. On the other hand, if humans understand their own role as the creators of their world, that's a much more healthy point of view.

It raises the dignity of humans. Indeed, if we do establish a new branch of human civilization on Mars that grows in time and potency to the point where it cannot really settle Mars, but transforms Mars, and brings life to Mars, we will prove to everyone and for all time the precious and positive nature of the human species and every member of it.

About an Element of Human
Greatness—Homer Hickam

What a grand day that was when Alan Shepard climbed into his Mercury capsule attached to a Redstone rocket—the rocket which would ultimately grow into the greatest rocket ever built, the Saturn V Moon rocket.

When I was a boy in Coalwood, WV, my greatest hero, besides my father and my mother, was Huntsville's Dr. Wernher von Braun. After Sputnik was launched in October 1957, the newspapers we received in our little coal camp were filled with stories of how American scientists and engineers were desperately working to catch up in the space race.

It was as if the science fiction I had read all my life was coming true. Gradually, I became fascinated by the whole thing. I read every article I could find about the men who built rockets and launched them, and kept myself pinned to the television set for the latest on what they were doing. Dr. von Braun's name was mentioned often.

At night before I went to sleep, I thought about what he might be doing at that very moment and imagined that he was down at Huntsville or Cape Canaveral high on a gantry, lying on his back like Michelangelo, working with a wrench on the fuel lines of one of his rockets.

I started to think about what an adventure it would be to work for him, helping him to build rockets and launching them into space. For all I knew, a man with that much conviction might even form an expedition into space, like Lewis and Clark. Either way, I wanted to be part of his team.

To do that, I knew I would have to prepare myself in some way, get some special knowledge about something. I was kind of vague on what it would be, but I could at least see I would need to be like the heroes in my science-fiction books—brave and knowing more than the next man. I started, in fact, to see myself out of Coalwood, to see a different future than might have been supposed for a coal mine superintendent's son.

To get an early start, five other coal miner's sons and I began to build our own rockets. The people of Coalwood called us, with some derision, the Rocket Boys. Years later, I would write a book[1] about those days, and later that book would get turned into a major motion picture called *October Sky*. And who would have guessed such a thing would happen because we coal miner's sons wanted to build rockets?

In the manner of Dr. von Braun, we Rocket Boys persevered until we were flying our rockets miles into the sky. Eventually, we boys of Coalwood won a gold medal at the National Science Fair, bringing unexpected honor to our little town and our high school. A lot of people said we were lucky. But it wasn't luck that allowed us to reach our goals. It was because we were prepared to take advantage of our successes, just as our heroes in Huntsville.

In the fall of 1957, Huntsville's Army Ballistic Missile Agency had just completed a flurry of advances in the field of rocket science. Under Dr. von Braun's guidance, it had solved most of the difficult problems associated with climbing out of

1. Homer Hickam, *Rocket Boys: A Memoir* (Doubleday Dell: 1998).

the gravity well into space. Everything was ready for a great leap forward, not because there had been a well-funded program for that purpose, but because the stubborn von Braun team had done everything they could legally, and sometimes a bit across the line, to prepare us for space.

When Sputnik hit this country like a flying sledge hammer, President Eisenhower turned to Huntsville to recapture a little American glory. Von Braun and his team launched our first satellite in sixty days of the go-ahead, and, less than a year later, they were cutting metal on another project even though it hadn't been officially approved—the Saturn family of rockets that eventually took the United States to the Moon.

This, I think, is history worth savoring and emulating. In many ways, the situation in space development is similar now to what it was in the years prior to 1957. There is confusion on what should be done in space and who should do it. Still, just as the Redstone was there to allow us to build the giant chemical rocket engines that made the Saturn rockets, the tools are available to us today, if only we have the courage to pick them up and use them for what we need to do. I want to give you a couple of things to think about. The first thing is that NASA is essentially a subversive organization. Anything else you hear is a public relations lie.

Some people, of course, will argue that there has never been a more conservative bureaucracy on this planet. NASA even takes polls to see what it should do next. It lets politicians decide technical matters.

I grant you NASA as a bureaucracy is timid. But what I mean is its charter. NASA is supposed to develop the means to allow American citizens to leave the planet. Leave! What could be more seditious than that?

I believe, in fact, that most NASA engineers and scientists are essentially subversives. It's in their psychological makeup. We strive for something greater than ourselves, something that is considered by most of the population of the rest of the world as being outlandish. What we want is to build machines that will literally allow people to leave this planet, and its governments, and its philosophies, and its religions behind to find a new world, new ways of governing themselves, new religions.

Back when I was a Rocket Boy in Coalwood, and in some trouble, as I almost always was, a preacher whom I adored—his name was the Reverend Little Richard of the Mudhole Church of Distinct Christianity—came to me and told me he had had a dream. He had seen men on the Moon, and I was one of them. When he woke, he had opened his Bible, and his eyes had fallen on the testament of Peter. "Nevertheless we, according to His promise, look for new heavens and a new Earth, wherein dwelleth righteousness," he quoted.[2]

It took me awhile to figure out what the old preacher was getting at, but then I understood it. I was a born subversive like St. Peter. My life was going to be dedicated to fulfilling His promise, the opportunity for mankind to look for those new heavens and new Earths, and all that would follow.

2. 2 Peter, 3:13.

Recalling the Reverend Richard's Bible verse, I believe it is the manifest destiny of this country to make all of us on the planet into a two-world species; our second world being our Moon. But I don't believe we will use chemical rockets to do it. It is time to put those old things in the museums where they belong. Let me explain.

It seems to me there are two ends of the rocket the American people love. The front end where the astronauts sit and the tail end where the rocket engines are bolted. It is time the working back end of the rocket got more emphasis than the front end. It is time that the old chemical rockets that have gotten us to this point be set aside and a new breed of big, bad rockets be constructed. If we don't do this, we are forever going to be stuck in low-Earth orbit—not doing any better than struggling to bolt together another Mir or Alpha or whatever you want to call it, a bunch of small modules like wieners in a string where astronauts can stay cooped-up for days, taking their blood pressures and peering at e-mail from home.

The Bush administration here in Washington is even now trying to figure out what to do with NASA. It's a problem. President George W. Bush knows, as all of us know when we are honest, that we are essentially spinning our wheels as far as spaceflight is concerned. And so it will always be until we start building advanced propulsion systems. Without them, we simply do not have enough lifting capacity or the acceleration to truly explore space. The one thing we don't need the ISS to teach us— we've already learned this—is that space is essentially bad for people. Zero-G is debilitating to our bones and muscles, and radiation from the solar wind and cosmic rays is likely to give us cancer if we're exposed to it over a period of time. To conquer space, we're going to need to get through it in a hurry inside spacecraft made of aluminum and steel and lead shielding. To go back to the Moon and definitely Mars with chemical rocket systems is asking for trouble, and that's trouble I don't think this Administration or the next or the next is ever going to take. So propulsion engineers, the good folks of Marshall Space Flight Center who head up propulsion for NASA, must take the lead, or we're never going anywhere.

Here's what my admittedly cloudy crystal ball says is coming: I believe we won't send humans back to the Moon or to Mars the way we did it in the Apollo program. Apollo pushed our technology to the outer edge of the envelope, and it took a brave corps of professional astronauts to go. That won't happen again.

I think spaceflight is going to evolve very much the same as the exploration of Antarctica. The first expeditions to the South Pole in the late nineteenth and early twentieth century were essentially sprints and publicity stunts, primarily accomplished for national prestige and personal glory, although often wrapped in the dubious veneer of science. These sprints were accomplished with the technology of the day—steamships to the ice shelf followed by the use of dogs and manpower to make a torturous journey to the pole and return. Many men died and many more suffered from frostbite, hunger, and exhaustion.

Roald Amundsen, a Norwegian, reached the Pole first on 14 December 1911, followed within days by his competitor, Britisher Robert Scott (who died with his entire team on the way back). After these highly adventurous and publicity-conscious expeditions, interest waned in duplicating their feats. It was far too expensive in money and blood to do something that had already been done. Amundsen and Scott's forays might be seen as roughly equivalent to the Apollo program.

It would be four decades later before the next people arrived at the South Pole. They were Americans, and they simply flew there in an airplane. I believe it will be in the space equivalent of this airplane that will carry people back to the Moon and to Mars and beyond.

But what kind of engine am I talking about that will power this space equivalent of the South Pole airplane? I believe the first one, if we're to see it in our lifetimes, must be a nuclear-fission

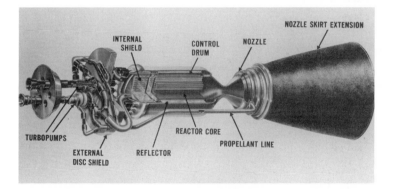

An explanatory drawing of the NERVA (Nuclear Engine for Rocket Vehicle Application) thermodynamic nuclear rocket engine. This program ran from 1960–1973 and was managed by NASA and the Atomic Energy Commission, but the nuclear engine was never actually deployed for a space mission. NASA Image NPO-70-15803.

rocket. For three decades, nuclear fission has been a dirty term in this country, which is really a shame. Despite the fact that nuclear propulsion is the best and safest way to fly major missions beyond Earth orbit, we stopped its development back in 1972. That was a terrible decision. We had at that point successfully tested nuclear rockets in the open air in Nevada—engines that could be operated with high thrusts for long durations, the key to the solar system. I believe it is time to go back to the future, to revisit the successful development of the old nuclear rocket, and to go forward with other designs, including the SAFE engine (Safe, Affordable Fission Engine) now being developed in Huntsville at Marshall Space Flight Center. Other similar engines are on the drawing boards at Glenn and JPL. We need to get behind the Rocket Boys and Girls in Huntsville and Cleveland and Pasadena, and show the courage it will take to chase away the Chicken Littles who deride nuclear energy in all its manifestations as somehow inherently evil. Nuclear energy, in fact, is one of the cleanest forms of energy we've ever known, and it's time somebody stood up and made its case.

Need I remind one and all that right now there are hundreds of nuclear reactors tooling around in the world's oceans, propelling submarines and aircraft carriers? How many of our sailors glow in the dark? Not one of them. This is safe technology, and I am tired of the fools in our society who fear it and keep it from its potential.

I believe it's time to put chemical rockets on the backburner for NASA. We should go ahead and build the International Space Station for political reasons and then turn the whole dull business over to a contractor or the National Science Foundation and get back to NASA's charter of truly exploring and utilizing space by constructing, testing, and flying some big, bad rockets. NASA should also be ordered to cooperate with the Department of Energy to provide energy from space to an energy-hungry world. This is the key. If we are to go into space in a big way, we have to have a big reason. That reason, I believe, will be the production of energy.

Right now, we're passing through a unique moment in time—a cheap energy bubble. This cheap energy bubble, actually available only in the Western world, is what allows us to transport ourselves, air condition ourselves, communicate with one another, run milling machines, make plastic, and do all the other things that make up our civilized, industrial, high-tech society. In this country, we can see that bubble just starting to burst in California. But it has actually already burst in most of the world.

The cheap energy bubble we enjoy is because of fossil fuel. When that goes away, and it will, if we have not taken the steps to prepare for alternative energy, our advanced society will collapse. Wind energy, ocean energy, land-based solar energy, all those can be added to the mix, but it will never supplant fossil fuel energy until we turn to space. In the meantime, however, while we work to come up with new forms of energy, we not only should be thinking about building nuclear plants for electrical power, but we also should start building nuclear engines needed to explore the solar system. Because those engines will bring us wealth, and energy, like we've never imagined. The solar system, if nothing else, is filled with energy, something we must have.

I am a boy of West Virginia, after all. I want to see a

Coalwood on the Moon or Mars. I want to read a memoir written by a boy whose father spends too much time mining helium-3 on the Moon or in the deep water mines of Mars. I want the solar system to become another place, a place of industry as Coalwood once was, a place where men and women can raise their families, and where a rough frontier—the roughest ever encountered by mankind—can be shoved back. And I want this country to lead the way.

One of the things Dr. von Braun instinctively knew was that it was as nearly important to explain why we should go into space as to build the machines to take us there. He wrote books and magazine articles, spoke to everyone he could formally and informally, and spent his life to his dying day seeking to educate everyone on the importance of and need for spaceflight.

I think all of us in this room need to continue that tradition, to get out into the public eye and explain why we should go and why we need to build the machines to go there.

Why should we go? I think there are actually two principle reasons:

1) Because the solar system is filled with cheap, clean energy, and we need to go get it, and, 2) this one may well be even more important, because we need a purpose for ourselves and our country.

Our Constitution established a relatively weak federal government and guaranteed strong individual freedoms. By doing so, what it really did was to allow the old Yankee Traders to go wherever they wanted to go, build up businesses, and generally outdo and outsell the rest of the known world hampered by kings and queens and imperial potentates.

That's how Americans got to be so rich. All those dollars and all the wealth that surrounds us didn't just happen because we're blessed. Our forefathers got out there and scratched for it using the freedom they had to do it.

You've heard of Yankee Traders. That's our heritage. We drive a hard bargain, and we make a profit. It's our way. Along the way, we just happen to bring our important values along with us—that of life, liberty, the pursuit of happiness, and the triumph of the individual over the government, any government, even our own.

A country needs a purpose, especially this country. I think we need an eternal frontier to push up against, our purpose to conquer and settle it. The nice thing about doing that is the solar system is a very rich place, filled with not only mineral wealth but energy, a nearly inexhaustible supply. And if there's one thing this country and this planet desperately needs and is willing to pay for now and forever is energy. The solar system is where it is. We've got to go after it.

I believe, then, that Americans have both a self-interest and a patriotic duty to convince ourselves it is time to take another giant step into space. The way to do that is to do what Dr. von Braun did—cut metal and start flying. Success engenders success. If we start flying advanced propulsion drives, we will, I believe, energize the country and perhaps inspire a new generation of Rocket Boys and Girls.

I call on NASA today to lay out a fifteen-year program to produce a working advanced propulsion engine in space, and that engine should most probably be powered by nuclear fission. Our elected representatives and the leaders of NASA should, in concert, move immediately to put this engine in NASA's budget with a fixed schedule for us to build it. If we can just get our first big, bad rocket booming around space, I believe there will be no holding Americans back from the new frontier.

In *Rocket Boys*, I tell of the time when I asked my father what was the hardest thing he'd ever learned. He leaned on the rail outside his office and said, "entropy," and then he explained. Entropy, he said, is the tendency of everything to move toward confusion and disorder as time passes. I was only fourteen at the time, so I'm certain I looked blank. "No matter how perfect the thing," he explained patiently, "the moment it's created, it begins to be destroyed." I asked him why that was so hard to learn. "Because," he smiled, "I don't want it to be true. I hate that it's true. I just can't imagine," he concluded, heading back inside his office, "what God was thinking."

Entropy ultimately killed Coalwood. When the coal was gone, everything fell into disrepair, and my father's hopes for the town crumbled. I don't want that to happen to my country, but it could, and I believe it will if we don't develop the solar system, open up a new frontier.

Dr. von Braun's grand dream need never die. NASA can spark a twenty-first-century revolution in transportation and energy that could fundamentally change the way we fly through space, power the world, even care for the sick if we do it right. All those things require cheap and clean energy. It exists in the solar system in a variety of forms. Solar energy is the most obvious form, but there are others, including helium-3, which may be the perfect fuel for fusion reactors, and also just happens to cover the Moon. I believe we must go after it. We just need to believe in ourselves and our purpose. I call on all of you here today to join together, get things moving again, and get serious about conquering space. If we do, we'll assure the country's prosperity, and the world's survival, too, for centuries. Along the way, maybe we'll finally understand why my old preacher thought St. Peter's comment about looking for new heavens was so important.

According to His promise, it's a big challenge, a huge responsibility, to try to fulfill such a promise and prophecy. I believe, however, the Rocket Boys and Girls of today are up to it.

So let's do it! Let's go!

The Ethics of Human Spaceflight—Laurie Zoloth

Left: Meriwether Lewis. This 1816 aquatint is by William Strickland. Image courtesy of the National Portrait Gallery. Image NPG.76.22.
Right: William Clark. This 1832 oil painting is by George Catlin. Image courtesy of the National Portrait Gallery. Image NPG.71.36.

There is a letter from Meriwether Lewis, struggling with the technology of the new collapsible iron-framed boat he has engineered for his great journey west into the Louisiana Territory. In this, we hear a story so familiar that it startles us. He is about to embark into a land so unknown that the maps are blank beneath his hands. He is inventing a technology that he and his mentor and master engineer President Thomas Jefferson have sketched out on an oak table at Monticello, and he is jerry-rigging, piece by piece, wood for iron. He is looking to hire on men for the journey who will be bold, physically able, and yet composed enough to live together in a small boat in terrifying danger. It is a ship that will fail him utterly in the middle of the journey. In the letter, he anguishes about funding for the project, about the time it is taking to make the boat, and about the way the trip must be timed precisely or postponed for another year. Congress is uneasy about the expenditures, and he must balance his work and his mission—commerce, science, exploration, and foreign policy.

"I visit him every day and endeavor by every means in my power to hasten completion of the work . . ." he says of the shipbuilder, and, of the river, its level dropping, he promises "this may impede my progress, but shall not prevent my proceeding, being determined to go forward . . . ."[1]

Jefferson replies, assuring him of the need for the mission, grounding his encouragement in the hopes that both the academic

1. See, for example, Stephen E. Ambrose, *Undaunted Courage: Meriwether Lewis, Thomas Jefferson, and the Opening of the American West* (New York: Simon and Schuster, 1996), p. 86.

science discovered and the social, agricultural, and entrepreneurial use of the new land justifies the difficulties. He writes of something more—of the intrinsic nature of the quest itself and of the obligations to the frontier borne by societies that encounter it. And while there was much to say then—and much criticism was given by contemporaries like Adams and others—there is still much to say now about the ethics of such an encounter. The arguments, the promises, and the vision that animated that journey are familiar because they are the substance of the vision that has animated much of NASA's efforts.

In this reflection, taken at the fortieth anniversary of NASA's years of space exploration and the 198th anniversary of the summer that Lewis and Clark set out, I want to consider exploration of this sort as a complex moral gesture. In this, I hope to both celebrate forty years of space exploration and to mark the way ahead. Taking this particular gesture against all other possible tasks reflects not only on who we are, and who we intend to become, but also on what we hold in common, as humans, and as Americans.

Like all exploration, space travel is far more than an extended journey; it challenges us to stand like the diarist Lewis before an unknown continent and an unknown territory. That it was profoundly inhabited, alive with others is a part of the paradox that faced him, and of course might well face us, as he could not know, and we cannot know. In this, the journal raises three core philosophical questions: first, of the nature of the human self, and how we are shaped by such an enterprise; second, of the technical process and rules of the task itself; and third, of the consequences of our moral actions on the world that we enter, and, by mere entrance, alter forever.

This chapter intends to lay out some essential ethics questions that might frame the next step of space exploration. In this, I undertake two sorts of tasks. The first is to respond to the core ethic question: Is it ethical to travel in space? The second, assuming for the moment that I can convince you that the first premise can be justified, is to comment on what ethical challenges will face us there.

It is appropriate to have a philosopher comment on this at the fortieth anniversary celebration, since it was also in 1962 that the National Academy of Science first convened a panel of philosophers to comment on space travel. They asked at that time whether it was indeed a worthwhile pursuit to travel in space and what might be expected of such an effort. What is at stake in any such boundary crossing is how the changing of essential human perimeters changes our own moral status. Will such boundary crossing worsen our human condition, or will it enhance it? In this way, the geopolitical quest is then linked to the quest for ontology, Pisarro hunting for the fountain of youth, for gold, and for territory.

What follows are a series of ethical claims that link the problem of discovery in the larger world and the attendant ethical dilemmas of our explorations, as well as how this exploration alters our concepts of life on Earth. In this, the role of the ethicist is to function as both a skeptic and a stranger, aware of the optimism of science and the pessimism of philosophy.

1. First premises and original claims: Why is it ethical to explore space?

The answer to the question—why space travel—is first ontological. What does the ethical gesture make of us as a society and a species?

> **A. Moral agency: A critical ethical task will be the transmission of why we have undertaken the work and what it makes of us to do so.**

How is space travel a moral activity? Like every gesture we make in the world, the activities themselves will change how we think about ourselves. However, is this itself suggestive of a hubric sense of ourselves? Does the very placing of ourselves at the center of the narrative begin the consideration of the task unjustly? In 1971, Lewis White Beck considered space exploration in his presidential address to the American Philosophical Association. Beck had a skeptical view at first of space travel. For Beck, the problem was our ability to justly and patiently sustain exploration that creates the essential ethical challenges. "[Space travel] is so far beyond human reach that it is not worthwhile discussing at a sober philosophical cocktail party . . . . The technology required presents no insurmountable obstacles; what stands in the way of using it is human unimaginative and impatience, and the instability of human civilization."[2]

But Beck was incorrect in this assessment. At the fortieth year of our reach, we have a Space Station despite all odds and

2. Lewis White Beck, "Extraterrestrial Intelligent Life," presidential address delivered before the American Philosophical Association in New York City, 28 December 1971.

instabilities—in fact, a station that represents an elegance and cooperation remarkable in the face of other conflicts, a victory of international imagination. Choosing space travel is a choice for a variety of social practices—it will require us to think about essential questions, and they are framing questions, not only in science, but classically, in ethics. Questions such as:

Are we overreaching our boundaries and
human limits?
We will always tell the truth?
What does it mean to be free?
How can we face death nobly?
Will we use resources fairly?
Can we be trustworthy?
What do we owe to one another?
What do we owe to the future we cannot know?

For such questions, surely not new ones, Greek philosophers, such as Aristotle suggested the need to develop habits of character—excellences. One way to think of the answers is to name the virtues called out by such questions: humility, veracity, courage, justice, and fidelity, which I suggest might be the ethical principles of space exploration.

We are shaped not only by such principles but also by narratives and historical arguments, and it is interesting that the problems of space and our relationship to it marked the earliest debates in philosophy, about the relationship of the human to the universe. One of the key arguments begins in the Greek consideration of how we should regard space; are we alone? Lucretius begins by noting that Earth has no privileged position in the universe, and that, in fact, other places might support life.[3] There might be innumerable, plural worlds, which have inhabitants—some like us and some unlike. The argument for our uniqueness is an argument about both limits and nature; are we unique? Are other worlds possible, and what does this make of our self-understandings?

In the Middle Ages, and through the eighteenth century, it was commonplace to understand the universe as an infinitely plural one—the universe was full of life, and humans were not alone. The entire thrust of emerging science, most especially Darwin's work, was a part of this understanding of the relationship of organism, contingency, and environment. Beck's claim is that space travel interests us because it offers a second chance at coming to an unblemished, uncorrupted world, an idea that animated much of eighteenth-century travel as well. The possibility of the great new chance, of new social possibilities for a new land, is a central part of the American vision that allowed Lewis, Clark, and others their optimistic spirit.

B. Is space travel just?

Let us concede that space travel is in fact ethical and perhaps ontologically necessary. But a central question of ethics is not only whether the act is good for us as humans, or is an intrinsic good, but whether space travel is just in both premise

---

3. See, for example, Whitney J. Oates, ed. *The Stoic and Epicurean Philosophers: The Complete Extant Writings of Epicurus, Epictetus, Lucretius, and Marcus Aurelius* (New York: Random House, 1940), pp. 111–114.

and process. Is space travel worth what it will cost in fiscal and human terms? This ethical problem is the first one about which most Americans and our international partners are concerned. It is the same one that John Adams argued with Jefferson about: Is exploration an unjust and wasteful use of scarce resources, better spent on urgent tasks at home? How can we launch our intricitly designed nineteenth-century boats or our twenty-first-century beautiful starships over a landscape of despair, illness, poverty, the closing of hospitals, the need for new elementary schools, over the tensions of war?

For space travel to be just, argue the Europeans, it must attend to principles of justice, which include the principles of vulnerability and of solidarity.[4] Since such principles include attention to the situation of the disempowered in human societies, and the need to maintain a democratic discourse about resources held in common, and since the assumption is that societies must find purpose in bearing the burden of the vulnerable, ethical space exploration ought to have applications from the science developed therein to human health and social welfare.[5]

Space travel can be justified as fair if direct benefits can be accrued by its undertaking. To a great extent, this can be said to be the case. First, the experience of microgravity has been found to create effects similar to aging in humans. Studies of the molec-

4. European Principles of Bioethics, 2000.
5. A. J. W. Taylor, "Behavioural Science and Outer Space Research," Aerospace Medical Association, Washington, DC, 1989; Jeffrey Davis. "Medical Issues for a Mission to Mars," *Aviation, Space, and Environmental Medicine*, vol. 70, No. 2, February 1999; O. P. Kozerenko, et al. "Some Problems of Group Interaction in Prolonged Space Flights," HPEE, April 1999, vol. 4, No. 1, pp. 123–127; Nick Kanos, et al. "Psychosocial Issues in Space: Results from the Shuttle/MIR," *Gravitational and Space Biology Bulletin*, 13 (2) June 2001, p. 35–45.

ular biology of bone loss are an example of this genre of work. Research from the first forty years of space travel is beginning to allow innovative medical research on osteoporosis. Second, gravity is a critical factor in development, sensorial and neurological orientation, and balance. Study of perception, hearing, balance, as well as studies about how neurological development proceeds in microgravity, is also ongoing. Such health-related research clearly will be a strong part of NASA's future duty. Cross-over and dual-use medical research is also at stake, with proposals that would allow the development of electronic-sensing devices. But much of what is valid about basic research is premised on what we cannot know. Support for any genre of research is predicated on this understanding. While it is the case that every protocol that calls for the use of animals as research subjects, for example, insists on a social justification as a part of NASA policy, and, while the intent of research is the betterment of the human (and animal) condition, and the relief of suffering, the reality is that, in true science, we cannot know the results or count on their application prior to the research itself.[6] Healthcare is only one sort of science that would justify the efforts in space as ethical in this way. Other areas clearly include research on climate change and other Earth science research only accessible from the vantage of space.

6. See, for example, the NASA Ames Standard Animal Policy forms for research protocols, as available on the Internet.

A second major justification for space exploration as a just endeavor is, I would argue, that it makes human society more likely to be free and at peace with one another. The science and engineering that is required and is intrinsic to space travel lends itself well to international cooperation. While it has not been the case necessarily in the past, the future of space exploration is profoundly international in character and is one of the few places where the human species has the opportunity to see itself as a collective community, as Earthlings. Such an interpretation allows a precondition for solidarity. Whether we act on this or whether we do not will be key in the determination of the justice of space travel.

However, the claim that space science is a diversion rests on an idea about science research as entirely instrumental. In this view, science is a zero-sum game, and, if we fund research on space, we will not be funding research on cancer or malaria. This is, I believe, a false distinction. Human subjects are shaped by the interrogation itself, and it is not entirely certain who or what we would be if we turned from the task. We must, in fact, do as we have always done—both the work of discovery, basic science, and the necessary work of compassion. It is the future that asks this of us; it is our children who will ask for the next step. We are needed in advance, by the work of basic science, as much as we may need what will be the insights of basic science to survive. Forty years ago, it was President Kennedy who urged the nation toward space travel, yet the most critical words from Kennedy were not about the Moon. They were about obligation and how obligation creates the American self, about obligations and not entitlements, asking us "what you can do for your country." In the era of a larger vision, we might expand this to "ask what you can do for your world."

3. Second premise: The ethical problems of process can be met fairly. Is the process and method of space travel ethical? Is it safe?

a. "Nothing will ever be attempted if all possible objections must first be overcome." (Samuel Johnson)[7]

Hence, if my arguments are correct, space travel is not only ethical, but narrative in character. We send a person to space to bear both the narratives of historical stories and data from the experiments we envision back to us—we expect that the person will also be changed. But this creates ethical problems. First, how will the process of travel shape the person, and how much risk can we expect the person to bear? Let me say that in the consideration of the ethical issues encountered by the process of space travel itself, we are reminded that the nature of experimental science itself—and space exploration is, among other things, a vast human clinical experiment—is a process fraught with human frailty, loss, risk, and error. We forget this at our own peril.

A key consideration will be how we understand the role of the crew and our corresponding obligations to the crew. Are they

7. Gwin J. Kolb, ed. *Rasselas and Other Tales*, vol. XVI of the Yale Edition of the Works of Samuel Johnson (New Haven: Yale University Press, 1990).

understood, principally as human subject in a difficult experiment, as primary investigators in charge of experiments in harsh conditions, as public servants akin to other public servants, with a limited number of workplace rights, but a higher level of duties and risk, as, say, firemen, as soldiers with a duty to explore and defend themselves? Each decision on the role of the crew allows for a different understanding of our duties and their correlative rights to our resources. Space travel, even to Mars, will necessitate an intensity of isolation in a small room, in conditions of weightlessness, for years. The crew will need different relationships to one another and to the machines that they will depend on. Because it is a genre of human experimentation, it must be preceded by animal experimentation, necessitating the humane use of animals in research. Microgravity presents unique and challenging conditions for animal welfare, requiring the coordination, expertise, and consistency of national oversight committees directed entirely toward this effort. NASA is committed to such oversight by a system of well-established mechanisms to review animal and human protocols, Institution Review Boards, and a Flight Animal Care and Use Committee established for this purpose. The IRB and the ACUC are guided in all of their functions by a clear and consistent advocacy for research animals and for the scientific enterprise itself. A clear focus on the issues of spaceflight allows the development of a mature expertise in the development, design, oversight, and reflection upon the science.

But the classic problems of clinical medicine only begin the ethical dilemmas that surround the process of spaceflight. The machine-human interface needed in space creates interesting and intriguing ethical issues, and the issues that surround the convergence of technologies in nanocomputers, genetics, and artificial intelligence are profound. First, if nanocomputers can be developed, it would be useful to have ways to monitor the crew without undue constraint. If internal monitors are used, how much would we accept monitoring and regulation of crew behavior? Should we allow the monitoring of blood sugar or serotonin? Should the nanochips be set to release medication to lower blood pressure, to alleviate stress? Should sleep perimeters be regulated and sleep enforced by biochemical regulations?

If a machine makes decisions about flight plan or choices about emergency strategic options which could be programmed with accuracy, should the machine be overridden and in what circumstances? When should human judgment guide the mission, and when should computers be allowed to make critical choices? Should we, in the future, select for genetic phenotypes more precisely than is already the case—for, of course, it is accepted that physical traits will be at stake in how crews are selected. Should screening include genetic testing, and should it inform our decisions about leadership? In the future, should we seek to genetically modify humans to make it safer for them to undergo microgravity, in the way medications are used to mitigate the experience, allowing for slower rates of bone loss, for example?

For all such decisions about medical conditions, is informed consent adequate, or is the yearning for the chance at being on the crew so powerful that true informed consent is meaningless, since refusal might well mean losing one's place on the mission? Is the risk of space travel simply an unacceptable risk?

That space travel is dangerous is not unique to other tasks in science. If we understand the crew not as research subjects but as Principle Investigators on a complex science mission, then it is not unlike the risks undertaken by the explorers of Antarctica or volcanologists. Here again we find the historical precedent useful in the careful reflection on that exploration which was a fiercely competitive race to the Pole—not unlike the conditions faced by crews in space. We know the paradox of these conditions.

In an article in *Discover* magazine, the diary of Frederick Cook, Arctic explorer in the nineteenth century, is quoted: "We are as tired of each other's company as we are of the cold monotony of the black night . . . physically, mentally, and perhaps morally, then, we are depressed, and, from my past experience, I know this depression will increase."[8] *Discover* magazine makes the case that the journey is most imperiled by the conditions of enclosure and boredom. Yet the intensity did not always breed despair in the eighteenth century—in many diaries, interestingly enough, the opposite is the case. The closeness and intensity created intensely loyal moral communities, ones where comrades never abandoned the ill, and where, in Meriwether Lewis' case, the very conditions of the trip, dangerous, challenging, kept him from the suicidal despair that awaited him in the ordinary life when he returned. But in space, the one constant of human

8. Quoted in William Speed Weed, "Can We Go to Mars Without Going Crazy?" *Discover*, May 2001, p. 36.

existence—gravity—will be absent, and for this we have no long-term data. Radiation, fire risk, and the possibility that bacteria and virus grow differently or faster in space are part of the unknowable risks that will be faced. Systems will fail as surely as Meriwether Lewis' collapsible boat failed to float. Finally, like the explorers of that century, the crews in space will be isolated from our world—in contact, after a lag, via virtual connection, but unable to affect events. They will be faced with a wide range of choices about how to negotiate human relationships.

Ethical issues may well arise, as they did for the Arctic explorators when a member became catastrophically ill. Unlike the recent cases of contemporary Antarctic missions, there will be no option for return or for more than simple medical and surgical interventions. Critical illness, accidents, or death may well occur. The usual understanding of bioethics regards the medical subject of research or in clinical medicine as a moral agent with full autonomy. But the conditions of space travel render this concept absurd—there surely will be no completely autonomous decision-making in the space capsules. Each action will deeply and mortally affect the lives of the others.

For such a dilemma, normative guidelines need to be created based on an open, reflective process, one that invites democratic reflection on the events of the journey with a full account of the risks. Lewis and Clark returned home to crowds of citizens along the Mississippi. In fact, their progress was reported as front-page news. This was not only true for the U.S. citizens, but for the Indian nations they passed through—hundreds witnessed the journey. This open, frank, and disclosive model—

the deep sense in the nineteenth century that the quest belonged to each citizen and that the government was each community—marked their journey. It is a model for us.

b. The ethics of encounter.

Of all the ethical considerations that characterized exploration in the eighteenth and nineteenth centuries, the leading one was clearly how to encounter the others that inhabited the terrain—the "empty space" of the Americas, Africa, and Southeast Asia. In the encounter with the native populations, the explorers had to first decide the moral status of these peoples and whether these people had the same sensibilities, duties, and rights as Europeans. Locke, Hobbes, and others sought to understand whether conscience was a feature of native persons.[9] But the encounter itself was fraught with inequalities. As Jared Diamond notes, the Europeans had access to weaponry, technology, food sources, and healthcare stability (relative to the bacterial and virally naïve populations of the Americas).[10] Encounters across serious differences with life forms we cannot know will be a sterling ethical challenge, both ethically and philosophically. We cannot anticipate the range of possibilities, and much of the speculation is outside the range of ethics and into the range of science-fictional scenarios, which have done a credible job in this realm.

9. John Locke, *The Second Treatise on Government* (Prometheus Books, 1992).
10. Jared Diamond, *Germs, Guns, and Steel: The Fates of Human Societies* (New York: W. W. Norton and Company, 1997).

c. The ethics of absence.

A final ethical challenge will be the converse of the issues above. What if we encounter nothing we can detect, merely a vast loneliness, a lack, an absence? This silence will carry its own theological and ethical consequence. Will this be warrant for our species? For how long? What of the idea that we merely are deceived in our perception, that, for example, RNA-based life forms or extremophiles we cannot detect surround us?[11]

2. Final considerations about the consequences of our action—Is the telos of space travel ethical?

We are bound to think of how the moral gesture of space travel affects ourselves; however, taking the idea of plural possible universes seriously means taking not only our right seriously, it means foregrounding our ethical obligations seriously. We will have no way of knowing how we will contaminate other planets, but the Heisenberg principle reminds us that we will alter the other place irrevocably. In our explorations of the past, even with our best intentions, and often without careful reflection, our species has rendered the terrain uninhabitable. Even with the best of intentions, we will carry with us bacteria, DNA, trash, and our own bitter history. What we fear and what we respect will shape our interpretation of what we see. We will step on things, and we will take samples. If we wish to analyze and

---

11. Beck, pp. 105–107; Ernan McMillin, "Life and Intelligence Far From Earth: Formulating Theological Issues, Extraterrestrial Life and World View," pp. 151–173; S. J. Dick, *Plurality of Worlds* (Cambridge: Cambridge University Press, 1982); *Star Trek*, the series.

understand such samples, we will take them back, or we may do so inadvertently, which may of course introduce into our world similar contaminants, yet another ethical dilemma.

In thinking about this problem, I asked my children for their thoughts, since, after all, it was more their problem than mine. My nine-year-old was careful to consider this matter, yet understood it not as a matter of interest, but of obligation. "What if," he asked me, "we are needed? What if we are supposed to explore space for some reason that we do not know about?"

Indeed. It is a part of our hubris to imagine we are alone, but another error to imagine we might turn away from the task. Can it be said that humans have an obligation to explore, not only because of our needs, but because we might be needed?

Unlike the optimistic Jefferson, content to civilize and garden after centuries of conflict, we fear that we will spoil a fragile nature, or we will be unable to tame it. Jefferson and his generation saw what they could bring to the new terrain; they were called by what they understood as its unfinished nature. Our view of space reflects this struggle to define its ends. Is nature pristine, is it normative, good as it stands, or can it be used, understood, charted, even altered, a place to repair in the sense of making it habitable, the classic work of civilization and cultivation itself?

We carry the disturbing human tendency to contaminate to be sure, but we also carry ideas of justice, democracy, and human imagination. We can be aware of all that we bring, including a crew, how it functions, and of our commitments to diversity and freedom.

Premises and considerations for reflection:

In previous work to justify animal research in space, NASA created a policy for the ethical treatment of animals called The Sundowner Principles.[12] The underlying values that emerged from the discourse that surround the creation of these principles are useful for a consideration of what undergirds space science in general and which might guide the near future of space exploration—respect for life and welfare of subjects, the integrity of science, the search for social good, which implies a value of solidarity with the peoples of Earth, and a broad notion of public accountability for both the science and the ethical priorities in space. Such considerations ought to guide how we consider the difficult choices we must make to explore space.

The considerations:

1. Welfare and excellence of care of all of the crew and animal subjects under the control of the Agency is the first priority of each and every project design. Research subjects deserve our special consideration for two reasons: the first on the grounds of essential ecological and moral concerns, such as nonmaleficence and basic stewardship of the vulnerable, and the second on the grounds of research respect for all subjects. Nonmaleficence guides every intervention in research whenever we are asking to study the behaviors and bodies of subjects, and thus we need to care for their welfare in all respects. Since space-

---

12. Joseph Beiletiski and Laurie Zoloth, "The Sundowner Principles: Ethical Treatment and Ethical Norms in Animal Experimentation," speech to the American Society for Bioethics and Humanities, Philadelphia, 2000. See also Sundowner Principles, NACUC NASA protocols and forms, for example Ames Space Center.

flight creates unique and unknown stress, it is rational to ask for a higher level of scrutiny. The increased level of trust demanded means that we have a particular duty to advocate for human and animal subjects in the design and review of projects.

2. Scientific endeavors shape the perimeters of all of the work. NASA projects are guided by several rationale, but the experiment of space travel is one is which we maintain research equipoise and which crew are both explorers and research subjects in conditions that may well afford society with knowledge about physiology that creates benefits on Earth that create a competing moral appeal in the consideration of the potential for burdens in the research design.

It is the science that motivates the goals of the work. Good science cannot occur if the subjects are compromised, but, if there cannot be valid scientific experiments, then a critical purpose of space travel is lost. Much of the work of space science is directed toward the creation, manipulation, and mastery of the new terrain. Every aspect of space habitat, medicine, long- and short-term effects of altered gravity, of radiation, or other challenges will need both animal testing and human crew if this project of survival and mastery of the environment is to continue. It is the science that justifies the resource expenditures and allows for the safe exploration of space to proceed in the most intelligent and thoughtful way.

3. Space science and flight is unique in many ways. One of the significant ways that it is important is that space research is large, visible, and very publicly funded. However, unlike the many other large and publicly funded projects, the public has

historically engaged as an open and enthusiastic witness of the space exploration aspects of the endeavor. Space research is not simply research on unknown environments, or unknown questions—it is exploration of unknown physical territory and, hence, draws on the oldest American imperatives. Since 1802, with the publicly funded project to explore the Louisiana Purchase, Americans have expected all such research to be democratically debated, widely published, and publicly accountable. Lewis and Clark published their letters, including letters to their families, in national newspapers, and NASA faces a level of interest and stake holding that is precisely similar and it is appropriate, however difficult, and, in this way, the crew are also public servants. Space is a part of what we hold in common, our common stock in our future, and, hence, it is fundamentally shared and a matter of public discourse.[13] This is appropriate—it is how Americans expect exploration to be carried out in a democracy, an idea as old as the idea of democracy itself, from the Greeks—humility, veracity, courage, justice, and fidelity.

4. The discourse needs to be joined about how to make space free. By this I mean free in the oldest sense—of liberation from the narrow place of restraint and domination to the large arena of human possibilities that is bounded by a new sense of social order, by the reality of human community, and the need

---

13. See, for example Michael Walzer, *Spheres of Justice: A Defense of Pluralism and Equality* (Basic Books, 1990) or John Rawls, *A Theory of Justice* (Cambridge, MA: Harvard University Press, 1999).

for justice. It is the Biblical metaphor of liberation from the slavery of Pharaoh's Egypt, "the narrow place" to responsibilities of the exilic journey, and it is familiar and recurrent. Such a sense of a liberatory, possible, risky, and burdened journey mediated the consciousness of the American endeavor from the beginning of the American experiment. We were not merely restless, or curious, or grumpy, or cramped, we were out for justice, and for the New Israel, for the City on the Hill. How can space remain free in this way? It is a question at stake in the news of the week as we celebrate the fortieth anniversary of space travel—will space be linked to an arms race? Will space be for sale, open for expensive tourism? Can space exploration be a multinational project? How can we negotiate such difficult and contentious issues? Are we fully committed to a model that in foregrounding science allows for the collaboration competition at the heart of science?

5. The exploration must be protective of our planet and our universe. To be an ethical enterprise, the ecological aspects of the journey must be rendered with the utmost care. Reciprocity must undergird the scientific impulse, and humility, with the limits of our wisdom and the power of our reach, must temper all aspects of the task.

I think this is possible, which of course renders space travel a moral enterprise. It is also an "authentic creation," which, as Albert Camus believed, was our central legacy—creations are authentic if they will exist after us. How does a society look forward and assess what might be left behind? It often does so by looking historically at other points of decision, where multigenerational tasks are self-consciously begun.

Considering the debates about our moral duties toward such tasks in the first year of the second century of the common era, 2001, it is worth remembering the arguments of Rabbi Tarfon, of the beginning of the first. Tarfon is asked about the permissibility of turning away from a task that one cannot possibly hope to complete personally. Can one stop work on projects that we understand we cannot complete—intergenerational tasks, construction, world repair, or, ultimately, justice itself? He reminds us that the task is without measure, and it can never be finished, but neither can we turn away. The work is not ours to finish, but the work is not ours to ever refuse to begin.[14]

14. *Perkei Avot*, 2:20–2:21.

*Future Visions for Scientific Human
Exploration—James Garvin*

An artist's conception of a future base and astronauts on Mars. This NASA image was produced for NASA by Pat Rawlings.

Human exploration has always played a vital role within NASA, in spite of current perceptions that today it is adrift as a consequence of the resource challenges associated with construction and operation of the International Space Station (ISS). On the basis of the significance of human spaceflight within NASA's overall mission, periodic evaluation of its strategic position has been conducted by various groups, most recently exemplified by the recent Human Exploration and Development of Space Enterprise Strategic Plan. While such reports paint one potential future pathway, they are necessarily constrained by the ground rules and assumptions under which they are developed. An alternate approach, involving a small team of individuals selected as "brainstormers," has been ongoing within NASA for the past two years in an effort to capture a *vision* of a long-term future for human spaceflight not limited by nearer-term "point design" solutions. This paper describes the guiding principles and concepts developed by this team. It is not intended to represent an implementation plan, but rather one perspective on what could result as human beings extend their range of experience in spaceflight beyond today's beach-head of low-Earth orbit (LEO).

Exploration of unknown frontiers has captivated the human spirit since the dawn of time, and it has been suggested that this spirit embodied the settlement and ultimately the development of the American continent and culture. How this concept has been extended to the space environment has largely been the stuff of science fiction, with the exception of the remarkable voyages of human beings to the Moon as part of NASA's Apollo Program. It is humbling to note that it has been more than twenty-eight years since humans have broken the bounds of the Earth's gravity field and physically entered the deep-space environment. Contrary to the misconceptions of many, the human experience in the space environment beyond the protective shielding of the Earth's Van Allen radiation belts is limited in its entirety to the brief flights of Apollo 8, 10, 11–17. Only approximately 220 hours of extravehicular activity (EVA) time was conducted by the Apollo explorers who visited the lunar surface between 1969 and 1972. Given this sometimes forgotten legacy, one quickly recognizes the extreme challenges of safely moving human explorers "on site" within deep space. This has posed a fundamental challenge to NASA in spite of the spectacular and perhaps unrivalled success of the Apollo Program of thirty years ago.

NASA has recognized the challenges of human exploration of deep space (HEDS) and most recently formed a small team of scientists, systems engineers, project managers, and biomedical experts to develop a vision with a somewhat radical set of boundary conditions. The challenge given to this wholly internal NASA exploration team was to develop a *scientifically driven* vision, enabled by new technologies and resource-constrained to eventually be implemented in an incremental fashion, rather than requiring a large total investment at the outset. The requirement that human exploration in this vision be driven by science and how human beings within deep space can uniquely contribute to the furthering of scientific progress is unprecedented. While the human spaceflight activities planned (and underway) on the ISS are certainly linked to fundamental scientific challenges, the broader scientific community has argued that the overriding motivation for the ISS in the first place was not as the logical next step for conducting high-priority scientific investigations from the vantage point of space. For example, the scientific catalysts for the ISS were not linked directly to the driving science challenges articulated by NASA's Earth and Space Science Enterprises. Ultimately, of course, the experience with long-term human adaptation to the near-Earth space environment and microgravity . . . provided by the ISS is a necessary stepping stone for humans to re-enter the deep-space environment as scientific explorers.

The challenge of science-driven human exploration is to develop the traceability from the most imposing scientific questions to human "on site" activities that will dramatically increase the potential for major discoveries and progress. The links between what humans can potentially accomplish by "being there" versus what can be achieved with high-bandwidth telerobotic presence (i.e., vicarious presence on site by humans off site) can be simply articulated. Our existing experience with human activities in space suggests that human-based field studies, provided robotic adjuncts are available to offer assistance and other infrastructure, are uniquely "discovery oriented," making possible rapid progress because of dynamic in-the-field responses to the local environment. The Apollo human expeditions demonstrated a high degree of onsite responsiveness to the lunar geologic environment, allowing for nearly instantaneous adjustments, thereby improving the field sampling results. In addition, humans can serve as effective erectors and operators of sophisticated apparatus in complex field environments. The Apollo experience again demonstrated the value of human-based setup of field geophysical equipment that even today defies our purely robotic capabilities. Thus, deriving the activity breakdown structure that optimizes insertion of the human into the scientific process is a key element of the vision recommendation for science-driven human exploration. Today, humans explore deep-space locations such as Mars, asteroids, and beyond, vicariously here on Earth, with noteworthy success. However, to achieve the revolutionary breakthroughs that have punctuated the history of science since the dawn of the Space Age has always required humans as "the discoverers," as Daniel Boorstin contends in this book of the same name.[1] During Apollo 17, human explorers on the lunar surface discovered the "genesis rock," orange glass, and humans in space revamped the optically crippled Hubble Space Telescope to enable some of the greatest astronomical discoveries of all time. Science-driven human exploration is about developing the opportunities for such events, perhaps associated with challenging problems such as whether we can identify life beyond Earth within the universe.

At issue, however, is how to safely insert humans and the spaceflight systems required to allow humans to operate as they do best in the hostile environment of deep space. The first issue is minimizing the problems associated with human adaptation to

---

1. Daniel J. Boorstin, *The Discoverers: A History of Man's Search to Know His World and Himself* (Random House: 1983).

the most challenging aspects of deep space—space radiation and microgravity (or non-Earth gravity). One solution path is to develop technologies that allow for minimization of the exposure time of people to deep space, as was accomplished in Apollo. For a mission to the planet Mars, this might entail new technological solutions for in-space propulsion that would make possible time-minimized transfers to and from Mars. The problem of rapid, reliable in-space transportation is challenged by the celestial mechanics of moving in space and the so-called "rocket equation." To travel to Mars from Earth in less than the time fuel-minimizing trajectories allow (i.e., Hohmann transfers) requires an exponential increase in the amount of fuel. Thus, month-long transits would require a mass of fuel as large as the dry mass of the ISS, assuming the existence of continuous acceleration engines. This raises the largest technological stumbling block to moving humans on site as deep-space explorers—delivering the masses required for human spaceflight systems to LEO or other Earth orbital vantage points using the existing or projected fleet of Earth-to-orbit (ETO) launch vehicles. Without a return to Saturn V-class boosters or an alternate path, one cannot imagine emplacing the masses that would be required for any deep-space voyage without a prohibitive number of Shuttle-class launches. One futurist solution might involve mass launch systems that could be used to move the consumables, including fuel, water, food, and building materials, to LEO in pieces rather than launching integrated systems. This approach would necessitate the development of robotic assembly and fuel-storage systems in Earth orbit, but could provide for a natural separation of low-value cargo (e.g., fuel, water) and highly sensitive cargo (i.e., humans and their associated systems and science tools). Future mass launch possibilities, including innovative laser levitation systems, beamed energy approaches, and even giant sling-based systems, could deliver the insensitive cargoes out of the Earth's gravity well at costs ultimately less than $100/lb, opening up new launch industries and distributing the launch requirements. Future-generation reusable vehicles could then concentrate on high-value payloads (humans, etc.) only and not have to carry low-value but high-volume materials all as part of the same system.

The new vision that emerges challenges the existing paradigms associated with ETO, in-space propulsion, as well as the types of trajectories to be used. Today we can hardly imagine hyperbolic, non-Keplerian transfers from Earth to deep-space destinations due to the unimaginable fuel mass requirements and in-space propulsion system performance levels. However, concepts for fusion-based propulsion could, in theory, deliver human-class systems from Earth to Mars in as short as ten days, and innovative ETO solutions are under study that might someday launch 1000-kg. cargoes to LEO for less than $100/lb and store consumables there in depots. This type of distributed approach would obviously require concerted investment and development, but, in theory, it could facilitate shorter-duration scientific exploration missions for humans to destinations as distant as the asteroid belt, Mars, or near-Earth objects such as comets and asteroids.

Any vision for scientific exploration of deep space by humans must always confront the issue of human safety and accessibility to the environment to be explored. It may be worthwhile to send human explorers hundreds of millions of kilometers to Mars orbit only to teleoperate robotic explorers on the planet's surface, but that would ultimately diminish the science "discovery" potential of humans on site. Human adaptation to deep space requires shielding or other countermeasures associated with space radiation, variable gravity, psychological stress, closed life support, and telemedicine. Materials science breakthroughs, as well as more effective space power systems, may ultimately provide the technological catalysts required. Carbon nanotubes (CNT) may deliver amazing strength-to-weight possibilities, thereby enabling structures hundreds of times less massive and contributing to shielding for human voyagers. Quantum energy delivery space power systems could provide kilowatts to megawatts of reliable power to sustain human crews and to power surface-based exploration. For example, on Mars there is an emerging requirement to access the subsurface to regions where liquid water may be stored. Such regions may lay hundreds to thousands of meters below the surface and necessitate complex drilling operations, all of which would require abundant and sustained power, as well as human tending.

The vision for human exploration considered by the NASA brainstorming team identified several breakthrough technology areas for which order-of-magnitude improvements over the existing evolutionary pace of development will be required to cost-effectively send humans into deep space with scientific activities in mind. Point design solutions certainly exist today that suggest there may be other nearer-term implementation solutions. However, if the overarching aim is to enable humans to conduct science operations on site in deep space, then technological breakthroughs that will have near-term robotic benefits are clearly needed. Aside for a few breakthrough technologies, there exists a thematically organized set of evolutionary technologies that must be matured if sustained human spaceflight beyond LEO is to be achieved. These include closed life support, telemedicine, information technology (automated vehicle health and maintenance), power delivery systems, artificial gravity, EVA systems, and human-robotic adjuncts and associated investigative sensors.

The science-driven, technology-enabled vision, as described above, would delay ambitious human-based deep-space exploration until sufficient technological breakthroughs are in hand to follow the guiding principles and ground rules. As a case example, we will consider scientific exploration of Mars by human beings as embodied by this technology-rich vision. In this case, a cascade of robotic forerunners integrated with NASA's ongoing robotic Mars exploration program would demonstrate the surface technologies and develop the required knowledge base about Mars and its environment before humans would be inserted on the surface. Initial human-based Mars exploration would be tactical, making use of the wealth of reconnaissance from precursor missions to focus on a few key sites at local scales with very goal-driven field exploration strategies. Initial tactical visits would be limited to six to nine months roundtrip, enabled by a new generation of in-space transportation systems, and involve surface residence times of thirty to forty days. These initial scientific expeditions might resemble the ongoing field exploration of the Earth's polar regions in which human explorers venture to a few key localities for very constrained periods of time but study the region from safe habitats telerobotically before conducting their optimized EVAs. First-visit human activities would be focused on highly informed sample collection, subsurface access, and on in-the-field-based discoveries. In addition, such initial human explorers would naturally serve as erectors of complex in situ geophysical and biogeochemical instruments and experiments which would operate long after the humans return to Earth. This sort of leave-behind infrastructure would be a vital part of the first wave of local, targeted human exploration, paving the way for longer-duration scientific outposts at the most promising sites. Initial visits would necessarily involve limited EVA activities, perhaps at the same level as those associated with the Apollo J-series missions (i.e., 7 hours per day). Conducting in situ life-detection experiments on possibly biologically related materials assuming appropriate safeguards may also emerge as key activities of the initial campaign of human visits.

The first wave of human expeditions would ultimately give way to a more sustained presence, with operations that could resemble those at Antarctic outposts here on Earth. Most importantly, as enabling technologies mature, human access to a variety of deep-space locations, including global access to the Martian surface, will be facilitated. Sending tactical, human onsite missions to targets where the scientific action has been identified by virtue of robotic precursors is an ultimate objective. In this vision, one can imagine a series of human-tended Martian surface drilling sites and associated astrobiology laboratories in which the chemical fingerprints of life are explored in situ, without the challenges of planetary protection (i.e., associated with returning samples of volatile materials to Earth safely, with no threats of backward contamination). Other possibilities might involve human field exploration of main-belt asteroids that could harbor evidence of ancient liquid water and potentially prebiotic indicators. Finally, there are scenarios in which human global access to the lunar surface would facilitate a series of sample return missions with which to determine the absolute chronology of the early-time portion of solar system history, and perhaps to link it with Earth, Mars, Mercury, and other objects.

This example vision might be viewed as an unwinding spiral of coupled scientific and technological developments that set the stage for an ever-increasing scope of human scientific activities at a wide variety of destinations in deep space. Discovering the limits of our science knowledge catalyzed by pushing the boundaries of our technological developments would offer a rich array of opportunities for engaging people, the ultimate customers of our deep-space exploration. This vision is all about using technology to dramatically amplify what we can learn scientifically and to facilitate a pace of discovery that provides excitement, adventure, and educational opportunities. By aggressively pursuing new technologies, the potential for feedback of such technologies into more traditional NASA programs, as well as to closer-to-home problems, would be maximized.

As with any vision, there are recognized implementation challenges. Moving human vehicles to Mars on time frames as short as weeks could require masses of fuel as large as asteroids, if one is not careful. Mass launch scenarios that require aggregation of hundreds to thousands of elements or fuel containers are almost unimaginable by today's standards. However, it is humbling to recall the lessons of history. In only twenty years after the voyages of Columbus and other early Renaissance explorers to America, Magellan and his team successfully circumnavigated planet Earth. Where could we be in twenty years with a sustained, integrated effort in which science, technology, and the adventure of human spaceflight work together? The possibilities are many, and what has been described here is one viewpoint developed during the course of an eighteen-month study by a single team of brainstormers. Our aim is to enable human exploration, and this vision is but one tale of a future that could be.

*Preparing for New
Challenges—William Shepherd*

William M. Shepherd, commander of the Expedition 1 crew to the International Space Station, secures his helmet to his Russian Sokol spacesuit during a training exercise on 20 October 2000. NASA Image JSC2000-E-27088.

This conference brings back fond memories for me. I recall Alan Shepard being at the Naval Academy in spring 1971 after he walked on the Moon. That was a special moment for me because I have always been interested in aviation. My dad was a Navy flier in World War II. My grandfather flew biplanes in France in World War I. It had long been one of my ambitions to be a naval aviator, but I found out shortly before that day in 1971 that I didn't have the eyesight to be a pilot. So I ended up being a Navy diver, a SEAL. I was sitting there listening to Al Shepard talk about his adventures on the Moon, and I was thinking I probably would never have to worry about doing anything like that. How strange events have turned.

I bring that up because I enjoy talking to kids and making education a very relevant part of space exploration. I think we often forget what impact exploration, technology, and human spaceflight have on the young kids of this country.

A little more personal history: I served in the Navy for thirteen years, was selected in 1984 to go to Houston and to start training as an astronaut. I flew three times on the Shuttle. They were very interesting flights. The longest flight I had was ten days. Right after that, I was asked to go to Washington for two weeks and help as the Administration had changed in 1992. In early 1993, there was a complete review of the Space Station program. We were in the middle of trying to decide whether Space Station Freedom would continue as a program or be canceled or be transformed into something else. So I had a lot of time in Washington working on how we would convert Freedom into something that was more feasible. At the end of that period, we got the Russians involved, which seemed politically to be a really good idea. One thing led to another, and, in 1995, I was asked to go to Russia and start training with another Russian cosmonaut, Sergei Krikalev, to be on the first human crew to the ISS.

So in early 1996, I started training in Russia and basically spent almost five years with Sergei and another Russian cosmonaut, Yuri Gidzenko, who is a colonel and a fighter pilot in the Russian Air Force. We trained, and, finally after four or five major delays, lots of slips and slides, we launched last October [2000] and flew what was called the first expedition to the ISS.

We launched from Baikonur on very much the same rocket that put Yuri Gagarin into space, pretty much an R-7 Russian booster. I was very impressed with the ride. It was very much like the Space Shuttle, very smooth. We lived on the International Space Station for 141 days. At the end of that period in March, the second expedition crew took over. A Russian and two American astronauts are up there now, Yuri Usachev, Susan Voss, and Jim Helms, and flying on the Space Station as we speak. So it was a real privilege to lead the charge, making the ISS a reality.

What I really want to talk about is not so much what we did, but what it all means. As the station is taking shape in orbit now, I think it might be easier to see not only what the issues were on its past development, but some of its future purpose.

To start off, it might be interesting to consider what differences and even similarities there would be between ISS as it is flying right now and a human expedition to Mars. So I'd like to touch on some of the ISS program lessons that are applicable. I think that a human expedition to Mars is going to be characterized by several broad themes.

First, it is going to be international. I do not believe that any single country has the financial resources, the technical know-how, or even the political will to carry out a large and costly exploration program to Mars or even back to the Moon. I think anything that we do in the future will be broad in scope and will be international in character, much as the partnership that's been formed to build and fly the station right now. The station partnership is certainly not perfect. It's got a lot of bumps on it, and we have had many tough times with our partners. Even today we're doing a lot of arguing, but the job is getting done, and I think it's a model for how we might do things in the future. It's not a perfect one, but it is functional.

Secondly, an expedition to Mars will be characterized by things that are big and heavy. I think that the vehicle that flies to Mars will be at least the size of the ISS and probably several times bigger. We need this because we have to have something that's very robust and self-sufficient. I think the crafts that we will see going to Mars are going to be too big to be put up by the current stock of expendable boosters or by the Space Shuttle. This is because we cannot afford to have a vehicle or a fleet of vehicles that are assembled in orbit that take thirty or forty launches. We have got to do this in a way that involves less risk and allows us to have better test and checkout on the ground as these large pieces are sent up into space so that the verification and integration job in orbit is not as large.

So risk, time, expense, and these checkout requirements are all going to favor putting large spacecraft on very large boosters and trying to get them into low-Earth orbit with a minimum number of launches. It is clear that we do not have that capability today. The expedition to Mars, regardless of how cleverly we design things, will be made up of a vehicle or a fleet of vehicles that will be assembled in orbit. Regardless of the technologies that we use, the vehicle that we would send to Mars would be very complex and will require the spacewalking and the robotics techniques that we're developing and proving right now on the ISS. There just is not any way to get some of the larger systems, particularly things like thermal systems or solar panel systems, completely checked out on the ground. We have to be able to assemble and perform such checkout procedures in orbit.

Third, expeditions to Mars also are going to need more power. We probably will not put humans on Mars transfer-trajectory using the energy for propulsion that would come from chemical sources. I am not sure that we will even be able to satisfy the ship's own energy needs with solar power or chemical reactions. Once we get to a place where we want to go, whether it is the Moon or Mars, we won't be able to do mining, in situ exploration of resources, efficient recovery, or manufacturing without large amounts of energy. We will need [not] only energy, but high power levels. To me, all these things say that the enabling technology to make this happen is going to involve some form of nuclear power.

Missions to Mars are going to need more speed. We need to cut down the transfer times to get humans and cargo to Mars and back. In order to do this, we have got to find propulsion schemes that are more efficient than the chemical rockets that we have now.

We are going to look for these types of engines that have to have both high thrust and high efficiency or high specific impulse.

Fourth, vehicles that go to Mars will be different from the vehicles that we fly today in space, because they will need to be highly autonomous. A Mars mission is going to have significant communications delays. Additionally, the trajectories to send vehicles to Mars do not lend themselves readily to good abort trajectories where we can get humans home quickly.

What this means is the crew onboard is going to have to manage and control what they are doing with not a lot of real-time help from the folks on the ground. It also means that the vehicles that we build are going to have to be survivable. We cannot put Soyuz capsules on this kind of vehicle and expect them to be able to get people home. We have to build spacecraft more like we build ships, where they can sustain some kind of damage from combat or some other catastrophe and still carry on. We cannot depend on mission control in either Houston or Moscow to keep turning things on and off all day on the station, because the awareness of what's happening on the ground is going to be 20, 30, 40 minutes behind what's really happening in space. All these factors indicate that the manner in which we are not only going to design but operate these vehicles is going to be quite different than what we are doing right now.

The last thing that I think is going to characterize a mission is that the environmental systems are going to look a lot like what you have on Space Station now. The Russians have done a pretty good job of closing off some of the environmental loops on their Zvezda module. Their equipment works very well and is very robust. It is a pretty well thought-out system. I think for missions up to six months, certainly a year, the kind of environment that we have right now on the ISS is a livable and workable place for humans.

If you include the other factors that I mentioned in trying to go to Mars and do exploration there, I think it starts to be within the grasp of what we could reasonably expect humans to be able to do. There are probably a lot of other questions to consider in relation to a human Mars exploration program, but I'll give you just some of the ones that I think about at night.

The big one would be: What should the structure of a large international partnership for exploration be? How will it be controlled? How will it be managed? I think the issues that we have had recently with the Russians as partners in the ISS of flying Mr. Tito show that we have not worked out all the bugs in the present International Space Station partnership about what partners are entitled to and how decisions are to be arrived at by consensus. We will have to get better at doing this.

Another question is how will high-energy density power plants and propulsors with high efficiency be developed? How will they be tested? Can we do this on orbit? How will a robust, reliable, maintainable, yet survivable spacecraft be designed, knowing that it diverges significantly from how we design here right now for work in space? How will the necessary political will be mobilized to carry out such a program?

One of my favorite questions relates not to technology per se, but to culture. Are we open to the necessary changes in our technical culture here in the United States, here at NASA, here in the ISS program, that will enable these other questions to be answered and these changes to be made? It is not just a question of adaptation at NASA or in the U.S. partnership for ISS, it's all the partners. It's the Russians, the Canadians, the Japanese, all the members of European Union, the Brazilians, and so forth. We have to do a much better job of trying to duplicate some of the efforts of the North Atlantic Treaty Organization to standardize military systems. We need standardized interfaces, components, and designs for space systems. We don't even have a coordinate system that everybody agrees upon as to how to lay a vehicle out and design it for space. We have a Russian system and a U.S. one, but we don't have a standard system.

We do not even have a standard vocabulary, let alone a standard understanding of even how we do the mathematics, the computer algorithms, the intricate computations that are necessary to guide vehicles through space. We think we understand each other's approach, speaking about the U.S. vis-à-vis Russia, but we have not converged on what we consider to be the optimal procedures. So I think something along the lines of the NATO experience is essential as a precursor to getting this exploration program done.

There are also some important issues regarding political will. I think the day is coming soon when people will be very interested to understand why the Space Station costs what it does. I don't think the question about Space Station or space exploration is really one of cost, but rather of value. People need to see a good return on what we've invested in space. They need to see that there is value present in what we are doing.

The ISS is hopefully going to do that in spades when we get more capability in orbit with our laboratories. But already you can see that commercialization of space is going to be a fundamental imperative. I think NASA is going to have to work hard soon at showing that the ISS has very strong commercial value.

I believe also that the role for astronauts, government astronauts, folks at NASA, my colleagues, is not really to do routine operations in low-Earth orbit anyway. We are the folks who need to go build the infrastructure. NASA should make plans to get out of the routine operation business in low-Earth orbit. We should let the marketplace drive that. As was the case in our own country with opening the West and laying rails coast to coast, the government helped facilitate putting down the infrastructure, and we let commerce proceed. I think space exploration will take place the same way. NASA needs to be the risk takers or the underwriters, and, once that era is over, we need to let commercialization proceed. Our business at NASA should be exploration.

So what are the problems? We've discussed many ideas. Some are well within reach of solution, while others are not. We've talked about commercialization. No one really knows well yet how to do it with the ISS, although we're heading in that direction.

We have talked about standards and the need for stronger ones, discussing how we merge all these partner nations into a joint international enterprise. I contend that we have done a particularly poor job of capturing the design philosophy. Why is the ISS or the Space Shuttle, or any rocket for that matter, built in its own specific way? Why didn't we put this filter over here? Why didn't we pick another type of fan to put in here? Why are the fuel lines or the fuel tanks for the rocket shaped in a particular way or made out of any kind of material? These aspects of development reside within U.S. contractors, but we are very poor at grasping the essence of why these decisions were made and cataloging them so people who come behind who have to do this job again can learn from these developments.

We talked a lot about having autonomous systems. We need to develop some type of pilot advocate, a high-level computer system that can logically run complex systems. But before that, probably the first step that we need to undertake is to decide on a common language, a common computer application, that will enable us to model complex systems to provide the necessary data for the human interfaces and displays, and to build and drive those displays.

We need to work on higher energy systems and nuclear power. I was over in Russia in 1995 at the Institute of Thermal Processes, where Sergei Korolev worked in the 1930s. We were over there looking at possibly adding a solar dynamic generator to the ISS. It was going to be a carbon block through which a gas flow led to a turban, and the carbon block was heated by a big parabolic reflector. The Russian engineers had a big chart on the wall, showing that this was going to be about a 10-kilowatt system with an exponential development growth. I said, "Can you guys say what the heat source is going to be for this system?" They said, "Well, of course. It's going to be nuclear." So I think the technical ability to do this is certainly right around the corner, and the Russians clearly have been thinking about this as well.

We also need to figure out how to energize a plasma by creating some type of electric-propulsion device that can both push very hard on an object with high thrust and also have very high exhaust velocity so it is very efficient. We are working on these types of technologies right as we speak in Houston at the Johnson Space Center. It is very possible that before too many years are out, this technology will be utilized to alleviate the ISS's drag. The ISS has a drag on it of only a couple of ounces, but this causes it to come down a couple hundred yards every day in its orbital decay. A highly efficient propulsion source could oppose that drag with a small amount of continuous thrust. This would keep the ISS from falling out of the sky and would significantly reduce our need to fly fuel up to re-boost it. This is important because when the ISS is fully assembled and weighs on the order of 500 tons, we're going to fly something on the order of 10 tons of propellant up to it every year to re-boost it. With a system such as this, that goes down by a factor of ten. We could have around a ton of hydrogen gas, but xenon might be a better gas that can be used in this plasma engine to propel the station and solve this re-boost requirement. This is a direct precursor of the kind of technology that will be useful to push crew transport vehicles to Mars.

There are a lot of other questions that have to be answered. We've got to go after closed life support. The Russians are working hard on this. They've got twenty years of experience base on Mir, and what they've put on the ISS right now is doing a good job of partially closing some of the gas and water loops. But we haven't totally solved such problems.

I think the biggest single issue in addressing any of the problems that I have mentioned is we do not have a way to marshal the adequate academic and intellectual resources in this country to solve these problems. So I would suggest that the country take a look at some of the national educational institutes that we have in the military. There are eight or nine of them— The National War College, Industrial College of the Armed Services, and so forth.

We need to have a National Space Institute that has some kind of Federal Charter. Its purpose would be to make available the intellectual resources necessary for the human exploration of space. It would have this as a single purpose. It would be a place where the appropriate knowledge, the experience, and the intellectual energy could be focused on this single goal. It would have the status of other national colleges. It would also be a virtual college or a university, and it would be collaborative with colleges and universities and other learning institutions all across the country and perhaps the world. Experts on space from almost any corridor could participate and contribute to what this institute would do.

It also would have very strong business participation. We would have folks from industry come to this environment, learn, go back and work, and then come back and teach. It would be a means for individuals to become more proficient in the technical engineering operations, as well as the business and political aspects of space exploration. Such a national space institute also would need to establish and maintain close contact with ongoing development and operations in human space programs. Now this needs not to be some intellectual outpost well away from what's happening, but rather needs to be in the center of the mix. I think this would probably be the most direct way to address the many technical, political, and cultural problems that I've mentioned.

In closing, my experience in space has led me to believe most strongly that all these issues can be addressed successfully. Sure, there are problems, but I think we are well on our way to finding the necessary answers, at least to enable human exploration back to the Moon and probably to Mars. It's just a question of marshalling the intellect and the will to go do it.

Should we have a clear policy at a national level to go make this happen? It is certainly something that can be carried out by the people who are in the field right now.

One hundred and forty-one days in space. I came back from my tour on the ISS thinking that spending your day looking at the surface of Earth is very enjoyable, a paradise as Mr. Tito says. Nevertheless, I really think that we should expand our vision by looking at the surface of other planets in this solar system. I am convinced that we have the means to do so.

About the Authors

Buzz Aldrin

Buzz Aldrin was one of the third group of astronauts named by NASA in October 1963. On 11 November 1966, he and Command Pilot James Lovell were launched into space in the Gemini 12 spacecraft on a four-day flight that brought the Gemini program to a successful close. Aldrin established a new record for extravehicular activity (EVA), spending 5 1/2 hours outside the spacecraft. He served as Lunar Module Pilot for Apollo 11, 16–24 July 1969, the first lunar-landing mission. Aldrin followed Neil Armstrong onto the lunar surface on 20 July 1969, completing a 2-hour and 15-minute lunar EVA. Buzz Aldrin was born 20 January 1930, in Montclair, NJ. He graduated from Montclair High School, Montclair, NJ; received a bachelor of science degree in 1951 from the United States Military Academy at West Point, NY, graduating third in his class; and received a doctorate of science in astronautics from Massachusetts Institute of Technology, Cambridge. His thesis was "Guidance for Manned Orbital Rendezvous." Aldrin has honorary degrees from six colleges and universities. Aldrin has received numerous decorations and awards, including the Presidential Medal for Freedom in 1969, the Robert J. Collier Trophy, the Robert H. Goddard Memorial Trophy, and the Hannon International Trophy in 1967. Since retiring from NASA in 1971 and the Air Force in 1972, he authored the auto-biography *Return to Earth* and several novels. Aldrin has remained at the forefront of efforts to ensure a continued leading role for America in human space exploration, advancing his lifelong commitment to venturing outward in space.

William Sims Bainbridge

William Sims Bainbridge received his doctorate in sociology from Harvard University in 1975, with a dissertation on the social movement that produced modern rocketry, published as The Spaceflight Revolution. Since then, he has published thirteen other books and well over a hundred shorter works, plus several computer software packages. His research tends to concentrate on the social science of technology or religion, and on employing surveying, observational, and historical methodologies. He currently works for an agency of the federal government, where he represents the social and behavioral sciences on the information technology and nanotechnology initiatives.

T. J. Creamer

Timothy J. Creamer graduated from Loyola College in May 1982 with a B.S. in chemistry. He was commissioned through the ROTC program as a Second Lieutenant in the U.S. Army. He entered the U.S. Army Aviation School in December 1982 and was designated as an Army Aviator in August 1983, graduating as the distinguished graduate from his class. He completed a master of science degree in physics at MIT in 1992. Creamer previously worked as a Space Operations Officer with the Army Space Command stationed in Houston, TX. Creamer was assigned to NASA at the Johnson Space Center in July 1995 as a Space Shuttle Vehicle Integration Test Engineer. Selected by NASA in June 1998, Creamer reported for Astronaut Candidate Training in August 1998. Having completed two years of intensive Space Shuttle and Space Station training, he has been assigned technical duties in the Space Station Branch of the Astronaut Office. Creamer was born on 15 November 1959 in Ft. Huachuca, AZ, but considers Upper Marlboro, MD to be his hometown.

Robert L. Crippen

Robert L. Crippen (Captain, USN, retired) recently retired from his position as president of Thiokol Propulsion. Prior to joining Thiokol, he served as a vice president with Lockheed Martin Information Systems in Orlando, FL. Crippen served as the Director of NASA's John F. Kennedy Space Center from January 1992 to January 1995. From January 1990 to January 1992 he served as Space Shuttle Director at NASA Headquarters in Washington, DC. Born on 11 September 1937, in Beaumont, TX, Crippen received a bachelor of science degree in aerospace engineering from the University of Texas in 1960. As a Navy pilot from June 1962 to November 1964, he completed a tour of duty aboard the aircraft carrier *U.S.S. Independence*. He later attended the USAF Aerospace Research Pilot School at Edwards Air Force Base, CA. Crippen became a NASA astronaut in September 1969. He was a member of the astronaut support crew for the Skylab 2, 3, and 4 missions, and for the Apollo-Soyuz Test Project mission. He was the Pilot of the first orbital test flight of the Shuttle program (STS-1, 12–14 April 1981) and was the Commander of three additional Shuttle flights. His accomplishments have earned him many notable awards—the NASA Exceptional Service Medal in 1972 and five awards in 1981, including the Department of Defense Distinguished Service Award, the American Astronautical Society of Flight

Achievement Award, the National Geographic Society's Gardiner Greene Hubbard Medal, and induction into the Aviation Hall of Fame. In 1982, he won the Federal Aviation Administration's Award for Distinguished Service, the Goddard Memorial Trophy, and the Harmon Trophy. In 1984, he received the U.S. Navy Distinguished Flying Cross and the Defense Meritorious Service Medal. He received NASA's Outstanding Leadership Medal in 1988 and three Distinguished Service Medals in 1985, 1988, and 1993. He is also a Fellow in the American Institute of Aeronautics and Astronautics, American Astronautical Society, and the Society of Experimental Test Pilots.

Stephen J. Garber

Stephen J. Garber works in the NASA History Office. He received a B.A. in politics from Brandeis University, a master's degree in public and international affairs from the University of Pittsburgh's Graduate School of Public and International Affairs, and a master's degree in science and technology studies from the Virginia Polytechnic Institute and State University. He has written on such aerospace history topics as the congressional cancellation of NASA's Search for Extraterrestrial Intelligence program, President Kennedy's attitudes toward space, the design of the Space Shuttle, and the Soviet space program.

James Garvin

James Garvin is the Chief Scientist for Mars Exploration at NASA Headquarters for the Space Science Enterprise. He came to NASA HQ in this position in April 2000 from NASA Goddard

Space Flight Center, where he had served as a Senior Earth and Space Scientist for fifteen years. Dr. Garvin was educated at Brown and Stanford Universities, receiving his Ph.D. in 1984 under professor Jim Head of Brown. His research focused on the sedimentary geology of the Viking Lander sites on Mars, as well as impact cratering processes. Upon arrival at NASA Goddard, Garvin spearheaded the development of planetary orbital laser altimetry as a tool for measuring landscapes on Mars and Earth. Garvin served as Chief Scientist for the Shuttle Laser Altimeter, which flew in Earth orbit in 1996 and 1997 on STS-72 and STS-85, respectively. Garvin has served NASA as a Science Team Member on the Mars Observer and Mars Global Surveyor missions, on RADARSAT (Canadian mission), and as the Project Scientist for the Earth System Science Pathfinders program. He has led NASA airborne remote-sensing missions to Iceland, the Azores, the Western U.S., Navassa Island, and other targets, many of which have served as analogues to Mars. More recently, Garvin has chaired the NASA Next Decadal Planning Team. Garvin looks forward to the day in which fusion propulsion will carry human explorers to the surface of Mars in little more than the time it took for the Apollo astronauts to reach the Moon.

Daniel S. Goldin

The Honorable Daniel S. Goldin became the Administrator of NASA in April 1992 and served until November 2001, making him the longest serving Administrator in NASA history. Before coming to NASA, Goldin was vice president and general manager of the TRW Space and Technology Group in Redondo Beach, CA.

During a twenty-five-year career at TRW, Goldin led projects for America's defense and conceptualized and managed production of advanced communication spacecraft, space technologies, and scientific instruments. He began his career at NASA Lewis Research Center, Cleveland, OH in 1962 and worked on electric propulsion systems for human interplanetary travel. Goldin is a member of the National Academy of Engineers and a Fellow of the American Institute of Aeronautics and Astronautics. He also received the Robert H. Goddard trophy from the National Space Club. He holds several honorary doctorates and other awards.

Frederick D. Gregory

Frederick D. Gregory is NASA's Deputy Administrator. Previously he served as Associate Administrator for the Office of Space Flight and as Associate Administrator for the Office of Safety and Mission Assurance. Mr. Gregory has extensive experience as an astronaut, test pilot, and manager of flight safety programs and launch support operations. As a NASA astronaut, he logged 455 hours in space—as Pilot for the STS-5 1B in 1985, as Commander aboard STS-33 in 1989, and as Commander aboard STS-44 in 1991. Mr. Gregory retired as a Colonel in the United States Air Force in December 1993. Mr. Gregory holds a bachelor of science degree from the United States Air Force Academy and a master's degree in information systems from George Washington University. He is on the board of directors for the Young Astronaut Council, Kaiser-Permanente, the Photonics Laboratory at Fisk University, and the Engineering College at Howard University. He is on the board of trustees at the Maryland Science Center, and he is a member of the executive committee of the Association of Space Explorers. His honors include the Defense Superior Service Medal; the Legion of Merit; the National Intelligence Medal of Achievement; two Distinguished Flying Crosses; sixteen Air Medals; the NASA Distinguished Service Medal; two NASA Outstanding Leadership Medals; National Society of Black Engineers Distinguished National Scientist Award; the George Washington University Distinguished Alumni Award; President's Medal, Charles R. Drew University of Medicine and Science; and honorary doctor of science degrees from the College of Aeronautics and the University of the District of Columbia. He also was awarded the Air Force Association Ira Eaker Award, as well as numerous civic and community honors.

Homer Hickam

Homer Hickam was born in 1943 and raised in the coal-mining town of Coalwood, WV, where he gained some fame as an amateur rocket builder. After leaving Coalwood, he earned a B.S. degree in industrial engineering from the Virginia Polytechnic Institute and then survived a combat tour in the Fourth Infantry Division in Vietnam. After his return from the war, Mr. Hickam found himself torn between his love of working on spaceflight and writing. He therefore decided to do both. While he worked for NASA designing spacecraft and training astronauts, he began a freelance writing career. In 1989, his first book, *Torpedo Junction*, a history of the battle against the U-boats along the American East Coast during World War II, was published. His next book was *Rocket Boys: A Memoir*, which was published in 1998 and became a number-one bestseller. It was successfully adapted into the motion picture *October Sky*. His next book, a techno thriller entitled *Back to the Moon* was published in the summer of 1999 and was quickly optioned by Hollywood. *The Coalwood Way*, a memoir which he calls an "equal, not a sequel" to *Rocket Boys*, was published in 2000 and quickly climbed the bestseller lists. The third book of his memoirs, entitled *Sky of Stone*, was published in October 2001.

John M. Logsdon

John M. Logsdon is the director of the Space Policy Institute of George Washington University's Elliott School of International Affairs, where he is also a professor of political science and international affairs, and director of the Center for International Science and Technology Policy. He holds a B.S. in physics from Xavier University and a Ph.D. in political science from New York University. He has been at George Washington University since 1970, and he previously taught at The Catholic University of America. He is also a faculty member of the International Space University. He is the author of *The Decision to Go to the Moon: Project Apollo and the National Interest* and general editor of the eight-volume series *Exploring the Unknown: Selected Documents in the History of the U.S. Civil Space Program*. Dr. Logsdon is Chair of the Advisory Council of the Planetary Society and a member of the Society's board. Dr. Logsdon has been a Fellow at the Woodrow Wilson International of the National Air and Space Museum. He is a Fellow of the

American Center for Scholars and was the first holder of the chair in the Space History Association for the Advancement of Science and of the American Institute of Aeronautics and Astronautics.

Charles Murray

Charles Murray is the coauthor with Catherine Bly Cox of *Apollo: The Race to the Moon* (1989). His other books include *Losing Ground: American Social Policy 1950–1980* (1984), *In Pursuit: Of Happiness and Good Government* (1988), and *The Bell Curve: Intelligence and Class Structure in American Life* (1994, with Richard J. Herrnstein). He is currently working on a new book entitled *Truth and Beauty: An Inquiry Into the Nature and Causes of Human Accomplishment*. In addition to his books and articles in technical journals, Dr. Murray has published extensively in the popular press. Dr. Murray has lectured at all of the leading American universities and at institutions throughout the world. He has been a witness before Congress and consultant to senior government officials of the United States, England, Eastern Europe, and the OECD. He has been a frequent guest on major network news and public affairs programs. Dr. Murray has been affiliated with the American Enterprise Institute since 1990. From 1981–1990, he was a Fellow with the Manhattan Institute, where he wrote *Losing Ground* and *In Pursuit*. From 1974–1981, he worked for the American Institutes for Research (AIR), one of the largest of the private social science research organizations, eventually becoming chief scientist. Before joining AIR, Dr. Murray spent six years in Thailand, first as a Peace Corps volunteer attached to the

Village Health program, then as a researcher in rural Thailand. Dr. Murray was born and raised in Newton, IO. He obtained a B.A. in history from Harvard and a Ph.D. in political science from the Massachusetts Institute of Technology.

William M. Shepherd

William M. Shepherd was selected as an astronaut candidate by NASA in May 1984. A veteran of four spaceflights, Shepherd has logged over 159 days in space. Most recently, he was the Commander of the Expedition-1 crew on the International Space Station (31 October 2000 to 21 March 2001). Earlier, he made three flights as a Mission Specialist on STS-27, STS-4 1, and STS-52. From March 1993 to January 1996, Shepherd was assigned to the Space Station Program and served in various management positions. William Shepherd was born 26 July 1949 in Oak Ridge, TN, but considers Babylon, NY his hometown. He graduated from Arcadia High School, Scottsdale, AZ in 1967, received a bachelor of science degree in aerospace engineering from the U.S. Naval Academy in 1971, and the degrees of ocean engineer and master of science in mechanical engineering from the Massachusetts Institute of Technology in 1978. Before joining NASA, Shepherd served with the Navy's Underwater Demolition Team Eleven, SEAL Teams One and Two, and Special Boat Unit Twenty. He is a member of the American Institute of Aeronautics and Astronautics (AIAA) and a recipient of NASA's "Steve Thorne" Aviation Award.

Asif A. Siddiqi

Asif A. Siddiqi was born in Dhaka, Bangladesh and educated in Bangladesh, the United Kingdom, and the United States. He obtained his B.S. in electrical engineering and M.S. in economics from Texas A&M University, and his M.B.A. from the University of Massachusetts-Amherst. He has published extensively in such journals as *Spaceflight*, the journal of the British Interplanetary Society; *Quest: The Journal of Spaceflight History*; and *Countdown* on the history of space exploration. He contributed to articles on the Russian aerospace industry for the year 2000 edition of *Encyclopedia Britannica*. Mr. Siddiqi was the winner in 1982 of a national essay contest on space exploration sponsored in Bangladesh by the United Nations. He was the 1997 recipient of the Robert H. Goddard Historical Essay Award sponsored by the National Space Club. In 1998, the American Institute for Aeronautics and Astronautics (AIAA) awarded the manuscript for his book *Challenge to Apollo* the prize for the best historical manuscript dealing with the science, technology, and/or impact of aeronautics and astronautics on society. He is a member of the British Interplanetary Society. He is currently a Ph.D. candidate in the department of history at Carnegie Mellon University in Pittsburgh, PA.

Neil de Grasse Tyson

Neil de Grasse Tyson was born and raised in New York City, where he was educated in public schools clear through his graduation from the Bronx High School of Science. Tyson went on to earn his B.A. in physics from Harvard and his Ph.D. in astrophysics from Columbia University. Tyson's professional research interests are varied, but they primarily address problems related to star formation models of dwarf galaxies, exploding stars, and the chemical evolution history of the Milky Way's galactic bulge. Tyson obtains his data from telescopes in California, New Mexico, Arizona, and in the Andes Mountains of Chile. In addition to dozens of professional publications, Dr. Tyson continues to write for the public. In January 1995, he became a monthly essayist for *Natural History* magazine under the title "Universe." Tyson's recent books include a memoir entitled *The Sky is Not the Limit: Adventures of an Urban Astrophysicist*, the companion book to the opening of the new Rose Center for Earth and Space; *One Universe: At Home in the Cosmos* (coauthored with Charles Liu and Robert Irion); and a playful Q&A book on the universe for all ages entitled *Just Visiting This Planet*. Tyson is the first occupant of the Frederick P. Rose Directorship of the Hayden Planetarium, and he is a visiting research scientist in astrophysics at Princeton University, where he also teaches.

Charles D. Walker

Charles D. Walker is a director of marketing for the Boeing Company's Space & Communications office in Washington, DC. In the early 1980s, he flew on three Space Shuttle missions as the first industry-sponsored engineer and researcher. His work as Payload Specialist astronaut on those missions included low-gravity purification of biomedical preparations and protein crystal growth. Programs on which Walker has worked include the Space Shuttle and the International Space Station. He has been involved in space systems engineering design, development, and operations planning through NASA, industry, and other organizations. Walker previously worked for the U.S. Navy in manufacturing and with the U.S. Forest Service in engineering and as a forest fire fighter. Walker serves on the boards, and as an officer, of numerous nonprofit and educational groups including the Challenger Centers for Space Science Education, the Association of Space Explorers, and the National Space Society. He received degrees in aeronautical and astronautical engineering from Purdue University.

Mary Ellen Weber

Mary Ellen Weber has most recently worked with a venture capital firm to identify promising areas of space research and related companies for investment. Previously she was the Legislative Affairs liaison at NASA Headquarters in Washington, DC. She has served on a team designated to assess and revamp the Space Station research facilities. Dr. Weber's principal technical assignments within the Astronaut Office have included Shuttle launch preparations at the Kennedy Space Center, payload and science development, and development of standards and methods for crew science training. Dr. Weber was selected by NASA in the fourteenth group of astronauts in 1992. Her most recent mission was STS-101, the third Shuttle mission devoted to International Space Station construction. Her first mission was STS-70; on this mission she performed biotechnology experiments, growing colon cancer tissues. Dr. Weber was born on 24 August 1962 in Cleveland, OH; Bedford Heights, OH is her hometown. She graduated from Bedford High School in 1980; received a bachelor of science degree in chemical engineering from Purdue University in 1984; and received a Ph.D. in physical chemistry from the University of California at Berkeley in 1988. She has received one patent and published eight papers in scientific journals. Dr. Weber has logged over 3,300 skydives since 1983. She was also in the world's largest freefall formation in 1996, with 297 people. In addition, she is an instrument-rated pilot.

Laurie Zoloth

Laurie Zoloth is a professor of ethics and director of the program in Jewish studies at San Francisco State University, and is the incoming president of the American Society for Bioethics and Humanities. In 2000, Professor Zoloth was a visiting scholar at the University of Virginia in the department of religion and the Center for Medical Ethics. She is also cofounder of the Ethics Practice, a group that has provided bioethics consultation and education services to healthcare providers and systems nationally, including the Kaiser Permanente System, five Bay Area medical centers, and regional long-term care networks. She is a bioethics consultant to NASA Ames Research Center and NASA's Interagency National Animal Care and Use Committees. She received her B.A. in women's studies and history from the University of California at Berkeley, her B.S.N. from the University of the State of New York, her M.A. in English from San Francisco State University, her M.A. in Jewish studies, and her Ph.D. in social ethics at the Graduate Theological Union in Berkeley. She has published extensively in the areas of ethics, family, feminist theory, religion and science, Jewish studies, and social policy, and has authored chapters in twenty-three books. Her book, *Health Care and The Ethics of Encounter*, on justice, health policy, and the ethics of community, was published in 1999. Her current research projects include work on both the ethics of ordinary life and emerging issues in medical and research genetics. In 2000, the National Institutes of Health awarded her an Ethical Legal and Social Issues of the Human Genome grant to explore the ethical issues after the mapping of the human genome.

Robert Zubrin

Robert Zubrin is the founder and president of the Mars Society, an international organization dedicated to furthering the exploration and settlement of Mars by both public and private means. He is also president of Pioneer Astronautics, an aerospace R&D company located in Lakewood, CO. Formerly a staff engineer at Lockheed Martin Astronautics in Denver, he holds a master's degree in aeronautics and astronautics, and a Ph.D. in nuclear engineering from the University of Washington. Zubrin is the inventor of several unique concepts for space propulsion and exploration, the author of over a hundred published technical and nontechnical papers in the field, and was a member of Lockheed Martin's "scenario development team," charged with developing broad new strategies for space exploration. In that capacity, he was responsible for developing the "Mars Direct" mission plan, a strategy which, by using Martian resources, allows a human Mars exploration program to be conducted at a cost 1/8th that previously estimated by NASA. He and his work have been the subject of much favorable press coverage. He has been featured in numerous television documentaries. Zubrin is also the author of the books *The Case for Mars: How We Shall Settle the Red Planet*, *Why We Must Enter Space: Creating a Spacefaring Civilization*, and *First Landing*, which was published by Ace Putnam in July 2001. Prior to his work in astronautics, Dr. Zubrin was employed in areas of thermonuclear fusion research, nuclear engineering, and radiation protection.

NASA History Series

Reference Works, NASA SP-4000:

Grimwood, James M. *Project Mercury: A Chronology*. NASA SP-4001, 1963.

Grimwood, James M., and C. Barton Hacker, with Peter J. Vorzimmer. *Project Gemini Technology and Operations: A Chronology*. NASA SP-4002, 1969.

Link, Mae Mills. *Space Medicine in Project Mercury*. NASA SP-4003, 1965.

Astronautics and Aeronautics, 1963: Chronology of Science, Technology, and Policy. NASA SP-4004, 1964.

Astronautics and Aeronautics, 1964: Chronology of Science, Technology, and Policy. NASA SP-4005, 1965.

Astronautics and Aeronautics, 1965: Chronology of Science, Technology, and Policy. NASA SP-4006, 1966.

Astronautics and Aeronautics, 1966: Chronology of Science, Technology, and Policy. NASA SP-4007, 1967.

Astronautics and Aeronautics, 1967: Chronology of Science, Technology, and Policy. NASA SP-4008, 1968.

Ertel, Ivan D., and Mary Louise Morse. *The Apollo Spacecraft: A Chronology, Volume I, Through November 7, 1962*. NASA SP-4009, 1969.

Morse, Mary Louise, and Jean Kernahan Bays. *The Apollo Spacecraft: A Chronology, Volume II, November 8, 1962–September 30, 1964*. NASA SP-4009, 1973.

Brooks, Courtney G., and Ivan D. Ertel. *The Apollo Spacecraft: A Chronology, Volume III, October 1, 1964–January 20, 1966*. NASA SP-4009, 1973.

Ertel, Ivan D., and Roland W. Newkirk, with Courtney G. Brooks. *The Apollo Spacecraft: A Chronology, Volume IV, January 21, 1966–July 13, 1974*. NASA SP-4009, 1978.

Astronautics and Aeronautics, 1968: Chronology of Science, Technology, and Policy. NASA SP-4010, 1969.

Newkirk, Roland W., and Ivan D. Ertel, with Courtney G. Brooks. *Skylab: A Chronology.* NASA SP-4011, 1977.

Van Nimmen, Jane, and Leonard C. Bruno, with Robert L. Rosholt. *NASA Historical Data Book, Volume I: NASA Resources, 1958–1968.* NASA SP-4012, 1976, rep. ed. 1988.

Ezell, Linda Neuman. *NASA Historical Data Book, Volume II: Programs and Projects, 1958–1968.* NASA SP-4012, 1988.

Ezell, Linda Neuman. *NASA Historical Data Book, Volume III: Programs and Projects, 1969–1978.* NASA SP-4012, 1988.

Gawdiak, Ihor Y., with Helen Fedor, compilers. *NASA Historical Data Book, Volume IV: NASA Resources, 1969–1978.* NASA SP-4012, 1994.

Rumerman, Judy A., compiler. *NASA Historical Data Book, 1979–1988: Volume V, NASA Launch Systems, Space Transportation, Human Spaceflight, and Space Science.* NASA SP-4012, 1999.

Rumerman, Judy A., compiler. *NASA Historical Data Book, Volume VI: NASA Space Applications, Aeronautics and Space Research and Technology, Tracking and Data Acquisition/Space Operations, Commercial Programs, and Resources, 1979–1988.* NASA SP-2000-4012, 2000.

Astronautics and Aeronautics, 1969: Chronology of Science, Technology, and Policy. NASA SP-4014, 1970.

Astronautics and Aeronautics, 1970: Chronology of Science, Technology, and Policy. NASA SP-4015, 1972.

Astronautics and Aeronautics, 1971: Chronology of Science, Technology, and Policy. NASA SP-4016, 1972.

Astronautics and Aeronautics, 1972: Chronology of Science, Technology, and Policy. NASA SP-4017, 1974.

Astronautics and Aeronautics, 1973: Chronology of Science, Technology, and Policy. NASA SP-4018, 1975.

Astronautics and Aeronautics, 1974: Chronology of Science, Technology, and Policy. NASA SP-4019, 1977.

Astronautics and Aeronautics, 1975: Chronology of Science, Technology, and Policy. NASA SP-4020, 1979.

Astronautics and Aeronautics, 1976: Chronology of Science, Technology, and Policy. NASA SP-4021, 1984.

Astronautics and Aeronautics, 1977: Chronology of Science, Technology, and Policy. NASA SP-4022, 1986.

Astronautics and Aeronautics, 1978: Chronology of Science, Technology, and Policy. NASA SP-4023, 1986.

Astronautics and Aeronautics, 1979–1984: Chronology of Science, Technology, and Policy. NASA SP-4024, 1988.

Astronautics and Aeronautics, 1985: Chronology of Science, Technology, and Policy. NASA SP-4025, 1990.

Noordung, Hermann. *The Problem of Space Travel: The Rocket Motor.* Edited by Ernst Stuhlinger and J. D. Hunley, with Jennifer Garland. NASA SP-4026, 1995.

Astronautics and Aeronautics, 1986–1990: A Chronology. NASA SP-4027, 1997.

Astronautics and Aeronautics, 1990–1995: A Chronology. NASA SP-2000-4028, 2000.

Management Histories, NASA SP-4100:

Rosholt, Robert L. *An Administrative History of NASA, 1958–1963.*
NASA SP-4101, 1966.

Levine, Arnold S. *Managing NASA in the Apollo Era.* NASA SP-4102, 1982.

Roland, Alex. *Model Research: The National Advisory Committee for
Aeronautics, 1915–1958.* NASA SP-4103, 1985.

Fries, Sylvia D. *NASA Engineers and the Age of Apollo.* NASA SP-4104, 1992.

Glennan, T. Keith. *The Birth of NASA: The Diary of T. Keith Glennan.*
J. D. Hunley, editor. NASA SP-4105, 1993.

Seamans, Robert C., Jr. *Aiming at Targets: The Autobiography of
Robert C. Seamans, Jr.* NASA SP-4106, 1996.

Project Histories, NASA SP-4200:

Swenson, Loyd S., Jr., James M. Grimwood, and Charles C. Alexander.
This New Ocean: A History of Project Mercury. NASA SP-4201, 1966;
rep. ed. 1998.

Green, Constance McLaughlin, and Milton Lomask. *Vanguard: A History.*
NASA SP-4202, 1970; rep. ed. Smithsonian Institution Press, 1971.

Hacker, Barton C., and James M. Grimwood. *On the Shoulders of Titans:
A History of Project Gemini.* NASA SP-4203, 1977.

Benson, Charles D., and William Barnaby Faherty. *Moonport: A History
of Apollo Launch Facilities and Operations.* NASA SP-4204, 1978.

Brooks, Courtney G., James M. Grimwood, and Loyd S. Swenson, Jr. *Chariots
for Apollo: A History of Manned Lunar Spacecraft.* NASA SP-4205, 1979.

Bilstein, Roger E. *Stages to Saturn: A Technological History of the Apollo/Saturn Launch Vehicles*. NASA SP-4206, 1980, rep. ed. 1997.

SP-4207 not published.

Compton, W. David, and Charles D. Benson. *Living and Working in Space: A History of Skylab*. NASA SP-4208, 1983.

Ezell, Edward Clinton, and Linda Neuman Ezell. *The Partnership: A History of the Apollo-Soyuz Test Project*. NASA SP-4209, 1978.

Hall, R. Cargill. *Lunar Impact: A History of Project Ranger*. NASA SP-4210, 1977.

Newell, Homer E. *Beyond the Atmosphere: Early Years of Space Science*. NASA SP-4211, 1980.

Ezell, Edward Clinton, and Linda Neuman Ezell. *On Mars: Exploration of the Red Planet, 1958–1978*. NASA SP-4212, 1984.

Pitts, John A. *The Human Factor: Biomedicine in the Manned Space Program to 1980*. NASA SP-4213, 1985.

Compton, W. David. *Where No Man Has Gone Before: A History of Apollo Lunar Exploration Missions*. NASA SP-4214, 1989.

Naugle, John E. *First Among Equals: The Selection of NASA Space Science Experiments*. NASA SP-4215, 1991.

Wallace, Lane E. *Airborne Trailblazer: Two Decades with NASA Langley's Boeing 737 Flying Laboratory*. NASA SP-4216, 1994.

Butrica, Andrew J., editor. *Beyond the Ionosphere: Fifty Years of Satellite Communication*. NASA SP-4217, 1997.

Butrica, Andrew J. *To See the Unseen: A History of Planetary Radar Astronomy*. NASA SP-4218, 1996.

Mack, Pamela E., editor. *From Engineering Science to Big Science: The NACA and NASA Collier Trophy Research Project Winners*. NASA SP-4219, 1998.

Reed, R. Dale, with Darlene Lister. *Wingless Flight: The Lifting Body Story*. NASA SP-4220, 1997.

Heppenheimer, T. A. *The Space Shuttle Decision: NASA's Search for a Reusable Space Vehicle*. NASA SP-4221, 1999.

Hunley, J. D., editor. *Toward Mach 2: The Douglas D-558 Program*. NASA SP-4222, 1999.

Swanson, Glen E., editor. *"Before this Decade is Out . . .": Personal Reflections on the Apollo Program*. NASA SP-4223, 1999.

Tomayko, James E. *Computers Take Flight: A History of NASA's Pioneering Digital Fly-by-Wire Project*. NASA SP-2000-4224, 2000.

Morgan, Clay. *Shuttle-Mir: The U.S. and Russia Share History's Highest Stage*. NASA SP-2001-4225, 2001.

Leary, William M. *"We Freeze to Please": A History of NASA's Icing Research Tunnel and the Quest for Flight Safety*. NASA SP-2002-4226, 2002.

Mudgway, Douglas J. *Uplink-Downlink: A History of the Deep Space Network 1957–1997*. NASA SP-2001-4227, 2001.

Center Histories, NASA SP-4300:

Rosenthal, Alfred. *Venture into Space: Early Years of Goddard Space Flight Center*. NASA SP-4301, 1985.

Hartman, Edwin P. *Adventures in Research: A History of Ames Research Center, 1940–1965*. NASA SP-4302, 1970.

Hallion, Richard P. _On the Frontier: Flight Research at Dryden, 1946–1981._ NASA SP-4303, 1984.

Muenger, Elizabeth A. _Searching the Horizon: A History of Ames Research Center, 1940–1976._ NASA SP-4304, 1985.

Hansen, James R. _Engineer in Charge: A History of the Langley Aeronautical Laboratory, 1917–1958._ NASA SP-4305, 1987.

Dawson, Virginia P. _Engines and Innovation: Lewis Laboratory and American Propulsion Technology._ NASA SP-4306, 1991.

Dethloff, Henry C. _"Suddenly Tomorrow Came . . ."_: _A History of the Johnson Space Center._ NASA SP-4307, 1993.

Hansen, James R. _Spaceflight Revolution: NASA Langley Research Center from Sputnik to Apollo._ NASA SP-4308, 1995.

Wallace, Lane E. _Flights of Discovery: 50 Years at the NASA Dryden Flight Research Center._ NASA SP-4309, 1996.

Herring, Mack R. _Way Station to Space: A History of the John C. Stennis Space Center._ NASA SP-4310, 1997.

Wallace, Harold D., Jr. _Wallops Station and the Creation of the American Space Program._ NASA SP-4311, 1997.

Wallace, Lane E. _Dreams, Hopes, Realities: NASA's Goddard Space Flight Center, The First Forty Years._ NASA SP-4312, 1999.

Dunar, Andrew J., and Stephen P. Waring. _Power to Explore: A History of the Marshall Space Flight Center._ NASA SP-4313, 1999.

Bugos, Glenn E. _Atmosphere of Freedom: Sixty Years at the NASA Ames Research Center._ NASA SP-2000-4314, 2000.

General Histories, NASA SP-4400:

Corliss, William R. *NASA Sounding Rockets, 1958–1968: A Historical Summary*. NASA SP-4401, 1971.

Wells, Helen T., Susan H. Whiteley, and Carrie Karegeannes. *Origins of NASA Names*. NASA SP-4402, 1976.

Anderson, Frank W., Jr. *Orders of Magnitude: A History of NACA and NASA, 1915–1980*. NASA SP-4403, 1981.

Sloop, John L. *Liquid Hydrogen as a Propulsion Fuel, 1945–1959*. NASA SP-4404, 1978.

Roland, Alex. *A Spacefaring People: Perspectives on Early Spaceflight*. NASA SP-4405, 1985.

Bilstein, Roger E. *Orders of Magnitude: A History of the NACA and NASA, 1915–1990*. NASA SP-4406, 1989.

Logsdon, John M., ed., with Linda J. Lear, Jannelle Warren-Findley, Ray A. Williamson, and Dwayne A. Day. *Exploring the Unknown: Selected Documents in the History of the U.S. Civil Space Program, Volume I, Organizing for Exploration*. NASA SP-4407, 1995.

Logsdon, John M., editor, with Dwayne A. Day and Roger D. Launius. *Exploring the Unknown: Selected Documents in the History of the U.S. Civil Space Program, Volume II, Relations with Other Organizations*. NASA SP-4407, 1996.

Logsdon, John M., ed., with Roger D. Launius, David H. Onkst, and Stephen J. Garber. *Exploring the Unknown: Selected Documents in the History of the U.S. Civil Space Program, Volume III, Using Space*. NASA SP-4407, 1998.

Logsdon, John M., gen. ed., with Ray A. Williamson, Roger D. Launius, Russell J. Acker, Stephen J. Garber, and Jonathan L. Friedman. *Exploring the Unknown: Selected Documents in the History of the U.S. Civil Space Program, Volume IV, Accessing Space*. NASA SP-4407, 1999.

Logsdon, John M., general editor, with Amy Paige Snyder, Roger D. Launius, Stephen J. Garber, and Regan Anne Newport. *Exploring the Unknown: Selected Documents in the History of the U.S. Civil Space Program, Volume V, Exploring the Cosmos*. NASA SP-2001-4407, 2001.

Siddiqi, Asif A. *Challenge to Apollo: The Soviet Union and the Space Race, 1945–1974*. NASA SP-2000-4408, 2000.

Monographs in Aerospace History, NASA SP-4500:

Launius, Roger D. and Aaron K. Gillette, comps. *Toward a History of the Space Shuttle: An Annotated Bibliography*. Monograph in Aerospace History, No. 1, 1992.

Launius, Roger D., and J. D. Hunley, comps. *An Annotated Bibliography of the Apollo Program*. Monograph in Aerospace History, No. 2, 1994.

Launius, Roger D. *Apollo: A Retrospective Analysis*. Monograph in Aerospace History, No. 3, 1994.

Hansen, James R. *Enchanted Rendezvous: John C. Houbolt and the Genesis of the Lunar-Orbit Rendezvous Concept*. Monograph in Aerospace History, No. 4, 1995.

Gorn, Michael H. Hugh L. *Dryden's Career in Aviation and Space*. Monograph in Aerospace History, No. 5, 1996.

Powers, Sheryll Goecke. *Women in Flight Research at NASA Dryden Flight Research Center, from 1946 to 1995*. Monograph in Aerospace History, No. 6, 1997.

Portree, David S. F. and Robert C. Trevino. *Walking to Olympus: An EVA Chronology*. Monograph in Aerospace History, No. 7, 1997.

Logsdon, John M., moderator. *Legislative Origins of the National Aeronautics and Space Act of 1958: Proceedings of an Oral History Workshop.* Monograph in Aerospace History, No. 8, 1998.

Rumerman, Judy A., comp. *U.S. Human Spaceflight, A Record of Achievement 1961–1998.* Monograph in Aerospace History, No. 9, 1998.

Portree, David S. F. *NASA's Origins and the Dawn of the Space Age.* Monograph in Aerospace History, No. 10, 1998.

Logsdon, John M. *Together in Orbit: The Origins of International Cooperation in the Space Station.* Monograph in Aerospace History, No. 11, 1998.

Phillips, W. Hewitt. *Journey in Aeronautical Research: A Career at NASA Langley Research Center.* Monograph in Aerospace History, No. 12, 1998.

Braslow, Albert L. *A History of Suction-Type Laminar-Flow Control with Emphasis on Flight Research.* Monograph in Aerospace History, No. 13, 1999.

Logsdon, John M., moderator. *Managing the Moon Program: Lessons Learned From Apollo.* Monograph in Aerospace History, No. 14, 1999.

Perminov, V. G. *The Difficult Road to Mars: A Brief History of Mars Exploration in the Soviet Union.* Monograph in Aerospace History, No. 15, 1999.

Tucker, Tom. *Touchdown: The Development of Propulsion Controlled Aircraft at NASA Dryden.* Monograph in Aerospace History, No. 16, 1999.

Maisel, Martin D., Demo J. Giulianetti, and Daniel C. Dugan. *The History of the XV-15 Tilt Rotor Research Aircraft: From Concept to Flight.* NASA SP-2000-4517, 2000.

Jenkins, Dennis R. *Hypersonics Before the Shuttle: A Concise History of the X-15 Research Airplane.* NASA SP-2000-4518, 2000.

Chambers, Joseph R. *Partners in Freedom: Contributions of the Langley Research Center to U.S. Military Aircraft in the 1990s.*

NASA SP-2000-4519, 2000.

Waltman, Gene L. *Black Magic and Gremlins: Analog Flight Simulations at NASA's Flight Research Center*. NASA SP-2000-4520, 2000.

Portree, David S. F. *Humans to Mars: Fifty Years of Mission Planning, 1950–2000*. NASA SP-2001-4521, 2001.

Thompson, Milton O., with J. D. Hunley. *Flight Research: Problems Encountered and What They Should Teach Us*. NASA SP-2000-4522, 2000.

Tucker, Tom. *The Eclipse Project*. NASA SP-2000-4523, 2000.

Siddiqi, Asif A. *Deep Space Chronicle: A Chronology of Deep Space and Planetary Probes, 1958–2000*. NASA SP-2002-4524, 2002.

Merlin, Peter W. *Mach 3+: NASA/USAF YF-12 Flight Research, 1969–1979*. NASA SP-2001-4525, 2001.

Renstrom, Arthur G. *Wilbur and Orville Wright: A Bibliography Commemorating the One-Hundredth Anniversary of the First Powered Flight on December 17, 1903*. NASA SP-2002-4527, 2002.